Outwitting SQUIRRELS

And Other Garden Pests and Nuisances

ANNE WAREHAM

MICHAEL O'MARA BOOKS LIMITED

To Charles: best pest

The original edition of *Outwitting Squirrels* written by Bill Adler was published in the United States by Chicago Review Press in 1982. This is a new version inspired by the original edition, published by Michael O'Mara Books and distributed in the United Kingdom.

First published in Great Britain in 2015 by
Michael O'Mara Books Limited
9 Lion Yard
Tremadoc Road
London SW4 7NQ

A CIP catalogue record for this book is available from the British Library.

Papers used by Michael O'Mara Books Limited are natural, recyclable products made from wood grown in sustainable forests. The manufacturing processes conform to the environmental regulations of the country of origin.

ISBN: 978-1-78243-370-5 in paperback print format
ISBN: 978-1-78243-371-2 in e-book format

3 4 5 6 7 8 9 10

Illustrations by Kate Charlesworth

Designed and typeset by K. DESIGN, Winscombe, Somerset

Printed and bound by CPI Group (UK) Ltd, Croydon, CR0 4YY

www.mombooks.com

Outwitting
SQUIRRELS

And Other Garden Pests and Nuisances

CONTENTS

Fighting infestations

Outwitting humans

INTRODUCTION

This is not the definitive text on nasties in the garden. This is, broadly speaking, my own experience of nasties in the garden and what I've learned about them, offered to you as honestly as possible. Which, you may agree, is a rare virtue in the gardening world, where blind optimism is generally favoured over absolute truth.

This should not make you downhearted. Yes, there are myriad pests and problems in the garden – one reason why I cannot be exhaustive here – but if you look around you, in most parts of the British Isles you will see thriving green places and spaces. Even the ones suffering from great neglect are likely growing something. Plants want to grow and they work hard at it; most of the time you need to get out of their way and let them get on with it.

However, if you decide you want to make trouble for yourself in the garden, there are several excellent routes to that end. The best is to grow fruit and vegetables. If you are wondering why you don't have issues with the bugs and diseases you hear many gardeners talk about, you are probably concentrating on growing flowers. The way to make real enemies in your garden is with edibles: everything, as well as you, will fancy having a bit.

The next best way to introduce yourself to a great load of nasties is to garden under glass. Bugs like living in snug and warm conditions with lots of good food to eat just as much as you do, and they may be harder to evict than your common human. But still, as good an idea as buying a greenhouse might seem, remember: just as there are nurseries full of thriving vegetables like French beans and peppers, there are also glasshouses full of unripe tomatoes and straggly geraniums. Something will always grow in there, if provided with a little life-giving water now and then, but to make the most of the benefits of a greenhouse, you must take extra care to combat your botanical tenants and get the environment just right.

And then you can make gardening hard work by being yourself. If you are tidy, a perfectionist or very competitive, you may find life in the garden that bit more challenging. Or, if you are living with someone with these traits, it will be both painful and noisy as you battle it out for dominance. It will be hard work if you feel yourself to be in a survival conflict with the hordes of pests that are intent on destroying your lovely garden. It will be even harder work if you have a neighbour, in-law or friend who has a tendency to treat your efforts with scorn or trample about without due care. In short, the greatest foes in a garden may be people.

So, look sceptically at all that 'Grow Your Own' and 'Here Comes the Allotment' fever that surges as an epidemic about every ten years – ten years being just about the length of time needed before people forget what being green-fingered is really like and fantasize about supermarket dependence again – because the main thing to remember is that being a gardener is hard, and not an exact science. It can also be hugely rewarding. You will either be lamenting hole-filled cabbages the size of golf balls or celebrating your prize-winning marrows. If you can accept that your new pursuit can be infuriatingly unpredictable, then you can also put up with a few pests and live a little easier.

Life in the garden will also be more straightforward if you can persuade yourself to be unlike 99 per cent of all British gardeners and be ruthless. Ruthless enough to throw out miserable plants that are not trying hard enough to win the race for exuberant, glorious life. Brave enough to change course if something is turning into far too much trouble and tough enough to experiment and risk being different. And if you are fortunate enough to plant something and see it grow, after nurturing it throughout the year, you must be strong enough to be able to cut it down again at the end of the season.

Do all of these things, and be up for the struggle, and you will enjoy an activity like no other. The fruits of your labours will be there to see, touch and taste and you will know that your success will have been down to your hard work and determination. It will also help if you're not too squeamish to pick up a slug (although, I am too squeamish for that and am still winning my own garden battles, so I think you can, too).

Anne Wareham

OUTWITTING CREATURES GREAT ...

Squirrels

*S*quirrels are like an intermittent mechanical fault of the worst kind. You know the sort of thing: your car starts stalling randomly, but if you take it to the garage, can you make it stall? Of course not. And the mechanics look at you sceptically and suck their teeth and clearly think you can't drive properly. Squirrels are similar in that when it comes to keeping them off your peanuts and birdseed, some guaranteed-to-defeat-squirrel feeders work like a dream for many people. But your squirrel will always be the one hanging upside down by one toe from a twig and eating the lot.

I have an anti-squirrel bird feeder with a wire mesh round it. I got it partly because it's quite attractive. I have yet to see a squirrel manage to obtain the nuts inside, or, indeed, try to. However, I know of one person who bought one who claims to have a picture of a squirrel happily ensconced inside the cage, devouring peanuts.

Before I got that one, I foolishly bought a complicated variety of feeder that abruptly shuts up shop when a squirrel lands on it. I had an annoying squirrel at the time, so it seemed like a good idea. I have never managed to assemble the horrible thing (the feeder, not the squirrel). It's still sitting half in and half out of its box, in bits, reproaching me. I'm not even sure how on earth I can dispose of it. Note to self: buy one I am likely to be able to understand how to construct next time.

Squirrels are great learners: this is their great survival trick. They can learn from another squirrel, or even from people, and they do it fast. It's not like teaching your dog to sit, so, sadly, if you have a family of squirrels around, it only takes one to learn the trick of breaking and entering and you can be sure that soon the whole family will be enthusiastic burglars. This ability to learn also clearly overcomes the fact that some squirrels aren't as bright as others.

They also have amazing memories, and it's said (I haven't investigated squirrels' brains in detail myself) that the hippocampus bit of their brain grows by as much as 15 per cent in the autumn when they are nut burying. This is the sort of effect learning 'the knowledge' has on the brain of a taxi driver, although I imagine that with the taxi driver it's less seasonal.

Squirrels are supposed to bury thousands of nuts in the autumn (who actually counted?). Whatever the exact number, it's certainly a lot, given the rate they scoff. Then they have to be able to find them again and some people claim they can make a three-dimensional map in order to do this. I feel mildly sceptical about these cartographical skills, but it's clear there would be no point in them burying nuts if they couldn't find them again when they need to eat. Considering human beings can't even find their car keys after they put them down only half an hour ago, it is seriously disturbing that squirrels can do so much better. Worse, if someone is watching a squirrel as it incarcerates a nut, it will simply pretend to bury the nut and then take it elsewhere, tucked in a cheek. Giggling as it goes, no doubt.

Perhaps their greatest skill, though, is sheer persistence, otherwise known as stubbornness. Bird feeder watchers will attest to this ability to try, try, try again. And then (yawn) try, try, try some more. And more. Until you feel satisfied you've defeated them this time and wander off for a cup of tea. An hour of trying later, the squirrel will be in. Maybe they just lack the human tendency to get bored fast.

So, if you are to try and outwit a squirrel, my first suggestion is to pick a dim and easily outwitted squirrel, not a grade-A squirrel with a First in Peanut Grabbing or a gold medal in the Squirrel Olympics. Then, bear in mind that when it comes to caging, most marauding mammals can get through a space the size of their heads. (Mice seem to manage to wriggle through even smaller ones – they have peculiar contracting skulls specially designed to make a mockery of you and of those people who tell you that you can make your house mouse-proof by filling in all the holes in your walls.) Having selected your squirrel, you can then think of ways to entertain it. For example, you can buy a bird feeder that, on detecting a squirrel, will tilt and spin. Sound like fun? Maybe not. Squirrels treat this feeder as an exciting new form of gym equipment as well as a source of free food. My worry about purchasing one of these is that, as it needs batteries, I feel sure my brightest squirrels will arrive just as they have run down or rusted. As the feeder spins, it may also fly open and shoot your birdseed out in a fine spray across the garden. So it may be wise to choose seeds you'd like to have sprout.

It could be that this indicates the best approach to squirrels: turn them into entertainment. You can create enormous obstacle courses that they will happily learn to navigate at breakneck speed. The trick is to introduce a new obstacle, wait for the squirrel to learn it and then add the next. Let your imagination invent impossible challenges and watch it conquer them. Your bird nuts have to become the reward, of course. All will be worth it when you invite your friends round for the Amazing Unprecedented Squirrel Acrobatics Feats Record. Though, truth to tell, the fun will be over in about twenty seconds, as squirrels are fast.

You may still prefer to properly protect your seed, however. So simple that even I could assemble it, the beautifully named 'baffle' is a plastic dome that fits on to the pole you hang your feeders on, protecting it from the route the squirrel will usually take to the feed – if you have the right kind of pole. Users report great success. The users, of course, being the squirrels.

The dumbest squirrels are, nonetheless, defeated. They approach in conventional squirrel style by climbing the pole (they might be better beaten by regular applications of grease on the pole, which could also offer you the entertainment of watching sliding squirrels) but then find they are stuck under the dome. The brighter squirrels, however, work out ways of taking flying leaps on to the top of the feeder pole above the not-so-baffling baffle. Their leaping abilities – honed as they jump around the trees of our woodlands – are quite phenomenal. Consequently you probably need a large and empty garden with your pole in the middle of it, far from fences, houses or any vertical structures. (Such an open space will enable your bird feeder to live up to its name, as the local birds of prey will now find the feeding birds a superb target and source of, well, bird food.) Those squirrels that are getting in form for their Olympic high jump will still manage to leap on to the baffle, probably finding it a comfortable picnic seat, thus rendering the design a success only part of the time.

A greased baffle may provide a slide, but the squirrel is unlikely to give up, and will generally soon break the baffle by its repeated heavy landings on to the one you've added above the nuts. I am not convinced that something with a wider surface area – such as a dustbin lid – as an alternative would exactly fit the bill as an attractive garden feature. But there is one option that could at least provide you with great entertainment: buy a cheap wireless doorbell, cover the bell part in plastic wrap and bury it in your peanuts. Wait. This is when, of course, the squirrel fails to appear. But when one does, press the doorbell ringer. You can get any number of different doorbell sounds to entertain you and the squirrels. Naturally, a gigantic bang would probably be best, but this is no doubt unfeasible. As is a remote-control detonator …

An imaginative manufacturer has added chilli to bird food on the assumption that squirrels won't eat it.

Remember when you didn't like vindaloo? You learned to like it; squirrels will learn to like it. It will then become their favourite.

Bird food eating may be the most unpopular thing squirrels do, but it's not the only one. They are multi-marauders whose palates are not restricted to nuts and seeds. We once had a fruit cage with raspberries, currants and other good fruit growing in it, safely tucked away from the birds. The squirrels happily chewed through the netting and had a feast, followed by the birds, who ate gratefully and greedily but then, unlike the squirrels, couldn't get out. Chasing birds round a cage full of cane fruit and vicious gooseberries is not a happy half hour.

Nor, in fact, was it much fun removing and replacing the netting roof every year against possible snowfall. We got more and more casual about it, as my husband was always sure that it wouldn't snow. And, of course, every year it did snow, so I would struggle with the nets alone while he was at work and snow was about to fall. Cold, wet and infuriating. So one year I just thought 'forget it', and the next day woke up to a totally demolished fruit cage – another pest preventative not recommended.

I started dreaming of an ornamental, beautiful and exotic cage, made entirely of unchewable wire, the perfect anti-squirrel fortress. But at times like these you have to ask yourself: just how much am I prepared to pay to save me having to go to the shops to get some raspberries?

Squirrels also dig holes in people's lawns, which are then helpfully enlarged by pet dogs. I think perhaps I am saved this because the squirrels don't recognize our cut grass as a lawn. They do regard tulips in pots as a buffet, though, and nuts, berries and fruit as a treat arranged by heaven especially for their delight.

They also enjoy a spot of bark stripping. Our land includes a small wood that once belonged to the Forestry Commission. It had previously grown conifers, which had been cut down a few years earlier, so that when we began looking after it we had a lot of saplings and young regenerated trees of an amazing variety – hornbeam, wych elm, beech, oak, sorbus, birch, cherry, willow, holly and yew – all this in just two acres. But they have been under attack from the moment they began to look like real trees (which takes a while), mostly from squirrels, which love to strip the bark in early summer to get at the sweet sap. They may well get type 2 diabetes from indulging in all that sugary stuff, which would serve them right, but this activity kills young trees and badly damages older ones. Sometimes, just for variety, squirrels will strip the bark off high enough for the tree to subsequently put out a new leader at a crazy angle. Understandably, the strain put on the trees is not good for them in the long term.

Squirrels are also reputed to raid songbirds' nests and eat the eggs and fledglings and, less horribly, they gnaw on hosepipes and plant labels. If they get into your house, bedlam ensues, together with an enormous insurance claim.

For all these things, we may wish them dead, or at least wish that they had an effective predator. In my experience, they do trap easily in a specially designed and baited squirrel trap, but unless you have a gun you'll have a demented live animal squawking at you and crashing around, with the problem of what on earth to do next. It's illegal to drown it (and also cruel) and impossible to bash it cleanly on the head without getting bitten. It is within the law to kill them, but causing your squirrel any unnecessary suffering will land you with a hefty fine, so any measure you take must be swift and humane. It is also illegal to try and palm it off on someone else by carting it off in a cage and releasing it a good distance away. Squirrels are legally vermin and you are not allowed to release them into the countryside. It's also futile in any case: I know someone who, having caught one, spray-painted it and then released it several miles away. It came back.

If you can't shoot a squirrel cleanly, you can only take it to a delighted vet, who I believe will be obliged to kill it for you at considerable expense because of it being verminous. The vet might not take kindly to being reminded of this obligation, though. And if you ever thought that cats could make you miserable when being transported to the vet, the squirrel will put them in the shade.

❧ WHAT TO DO ☙

Trap them if you can. It's up to you what to do next

Try a baffle – it's always worth a go. What makes one squirrel laugh will utterly confound another

Try greasing a pole (Vaseline or car grease) if the feeder is unleapable. Renew grease frequently

Have them for dinner if you can manage a kill – they make good eating and they're totally free range

Make friends with them

Study them for a PhD

❧ WHAT NOT TO DO ☙

Plant your bulbs with sprigs of gorse. If you've ever picked gorse, you'll understand

Surround all your plants with wire netting. It's not pretty

Get a cat – it will only join in with the game of persecuting the birds

Pee in strategic places to terrify the squirrels

Feed them

Rabbits

Rabbits are the sweetest little furry creatures, and a friend of mine who keeps them is very fond of hers. I'm sure I would be very polite to them if we met; I might even stroke them affectionately – it's very soothing to stroke a furry animal.

But as far as my garden is concerned, I'd rather see them dead.

For years, they gave me no trouble. I used to watch them come out to play at dusk in the neighbouring field. Then, suddenly, they invaded. And the worst part was the fact that we were still in the process of making the garden, and it's a large garden. Sadly, rabbits love to sample anything new. And to keep sampling until everything has been guzzled.

They will feed on ornamental plants, fruit and vegetables, especially young growers. You'll notice about half a metre's worth of leaves and shoots have been gnawed away (bark, too, if other food is scarce), as the blighters stand on their back legs to really get stuck in. I found myself walking around the garden looking for the telltale little black droppings, always hoping not to find any but stomping on the ones I did come across so I wouldn't mistake them for new ones next time round. I had fantasies that foxes or buzzards would catch them and eat them. Instead, I once had someone sitting eating their lunch in the garden who witnessed a buzzard dropping us a rabbit as a gift. Fortunately, the fall killed it, but I can't count on always being so lucky.

Garden-making can be a struggle at times, and I think one of my worst experiences was when I put out a lot of ornamental grasses that I'd grown from seed and which, after many transplants and lots of nurturing, had finally reached planting size. Finding them all gone the morning after was heartbreaking. And I think I had tried just about everything to deter the rabbits and to protect the plants.

For weeks, the garden reeked of Jeyes Fluid (see Legislation, page 165), and so did I. It's supposed to work as a rabbit deterrent, so I cheerfully sprayed it everywhere whenever I put out new plants. It didn't put the rabbits off. Nor, it seems, do any other smelly deterrents – at least, not consistently. It appears that noisome products put some rabbits off while others positively love them. When rabbits start to eat plants that have been soaked in repellent, you'd be inclined to think they view it as a condiment, but there'll always be someone with rabbits that do hate it and run away. Sometimes it seems they simply get accustomed to it. Personally, I think that unless it's 100 per cent effective 100 per cent of the time, I will dread going out there in the morning, and that's bad. And, given how random rabbit damage can be, who is going to spray everywhere in their garden after every downpour of wash-away rain?

I do also try protection. I used lots of little plant cages cut out of chicken wire, with satisfyingly sharp bits of metal left sticking up where you think a rabbit's chin might be if it dares to try and reach into the cage for a nibble. These work, but they are hard work: hard on the hands, hard on the tools you use for chopping up the chicken wire and hard on the appearance of the garden. But if the plant is small, you can cover it totally and, just to be sure, pin down the cage with tent pegs thrust through the holes so that single caged plant will be safe. (Don't forget, however, that slugs like new plants too, and slime their way over every bit of cage without a second thought.)

Don't imagine for one minute that there's any point in thinking of using rabbit-proof plants. Just as one rabbit will be deterred by a pong that is an appetizer to another, so one plant with a reputation for being totally and irrefutably rabbit proof will be manna from heaven to your particular plaguey rabbit. And a rabbit, it seems, will always try anything new once, even if it decides against a second bite. Never let a rabbit near a new plant.

You can go high-tech and attempt to deter rabbits with

nasty noises or even sprays of water. These machines have the same variable results as nasty smells but are much more expensive, so perhaps should only be employed after trying other methods first.

So, what's the answer? I'm sorry to tell you it's rabbit fencing, and even that is not 100 per cent secure. And, of course, it's expensive.

What does it involve? You need chicken wire with tiny holes, a minimum of 31 mm (1¼ in) so you can keep out the smallest rabbits. It needs to be 120 cm (about 4 ft) or more high. You may not wish to know this, but you *need* to know that rabbits can climb. They may be able to fly, too, but only in the dark when no one can see them, so we cannot be sure.

You are supposed to add high-tensile wire and the typical instructions for putting up your fencing imagine that you have perfectly flat land with nice soft soil for banging fence posts into. And, of course, that is as likely as flying rabbits.

I have undulating ground, parts of which are full of rocks. You should bury an apron of wire about 15 cm (6 in) wide in front of your fence so that rabbits planning to burrow underneath will encounter an underground obstacle before the one in front of them and give up. (They apparently don't then try burrowing from slightly further away. Or jump over.) This is not, frankly, always possible, but do your best. Also armour your gates, while trying not to dwell on that picture of Peter Rabbit crawling under one. Make sure they can't squeeze under by either lowering the gate as close to the ground as possible or by adding a useful strip of netting or a rubber flange along the bottom. Make sure people shut the gate in order for it to have some effect.

When my husband, Charles, rabbit fenced our perimeter, rather more messily than the neat descriptions specified, we discovered to our horror and chagrin that we had effectively fenced the rabbits in. Despair. We went away on holiday and some idiot came to the house and left the gate open. We arrived home to find it swinging in the breeze and imagined the garden would be full to busting with rabbits old and new, but no. They'd gone! This method of getting rid of captive rabbits cannot, however, be guaranteed.

So, having fenced, we have mostly been rabbit free, but never entirely. It seems that somehow a gate always gets left open, or a buzzard drops us a gift. Charles has to regularly check the fence to make sure it's still sound and that is a grim job for him, in among all the shrubs, brambles and rocks. Heavy snow is always an anxiety, too, as it operates like a lift, raising the height of the ground so rabbits may just step over if the snow drifts up against the fence.

So what do I do about these strays? Pray that they stay single, for a start. Or, even less likely, that they have taken a previously totally unheard of vow of celibacy, should more than one get in. I have a couple of rabbit traps and they are often effective – surprisingly, on several occasions I have caught rabbits in them with no bait in there at all. Or, confession time, Charles occasionally manages to shoot them if they sit still in a convenient spot long enough.

Rabbits that were romping round your garden will try very hard to re-enter if you manage to fence them out, but ones that have never known the delights of your special tasty veggies may not try to breach the fence to find them. Which sounds good, but doesn't explain how it is that we have a rabbit in the garden again.

❧ WHAT TO DO ☙

Save yourself multiple disappointments and just get on with the fencing

❧ WHAT NOT TO DO ☙

Waste time listening to most of the theories on bunny-repelling smells or rabbit-resistant plants

Cats

The worst things our cats used to do was to give rabbits a lift over the fence into the garden and to target fledgling birds. For days I would fondly watch birds flying into nests in holes in the hedge, only to then spot one of our cats enjoying the same sight. Any tiny bird trying out the flying idea would tragically end up in the cat's tummy. This is very hard to prevent if you have outdoor cats. It didn't stop there, however. I'd find our cats munching away at baby grass snakes and lizards, which I minded a little more than the voles and mice, although treading on a corpse when you open the bedroom door in the morning and step out without looking is a ghastly experience.

But even if you don't have cats, you may find them happily hop, skipping and jumping into your garden from elsewhere. You can't stop the whole neighbourhood from cat-keeping just because they stake out your bird feeder or use your seedbeds as a latrine. And these problems are not simply their lack of appeal to the fastidious, it's also the risk of adding toxoplasmosis to your nice healthy carrots.

As well as terminating local wildlife, their activities leave behind other telltale signs. Scorched plants can indicate where male cats have sprayed their urine to mark their territory, which they also sometimes do by scratching the bark of trees (I'd rather that than the armchair or wallpaper). They usually bury their poo in flower and vegetable beds by scraping soil over it, although in my experience they get lazy and only do half the job. Or they might leave it in the middle of the lawn, which is doubly unpleasant, as it seems as if they are taunting you.

There are many recommended ways of dealing with the cat problem, but, as usual, they have variable success rates and most have their associated problems. You might like the idea that moth balls will make a cat take one sniff and run away, but it is just as likely that it might eat one and die. Even if you could kid yourself that the worst thing in the world for a cat is the smell of coffee grounds, orange peel or lion poo, how long do you imagine these deterrents will go on providing a powerful pong, out there in the rain, sun and snow?

You can buy commercial versions of nasty smells, too, though why anyone would ever buy any having read the reviews is a mystery. Hearing about how cats like peeing on deterrent granules is not a confidence booster. I gather, though, that the smell is bad enough to be likely to keep you out of the garden, so perhaps you'll stop minding. Maybe that's how it's supposed to work?

You might have more success keeping them out of the garden with plastic spikes, supposing you have a convenient fence to nail the spikes on to. They would be difficult to apply to a hedge or a wall and you may find the colour as unpleasant as a cat. However, they have the added bonus of reducing the attraction of climbing over your fence to any other feral creature, including people. Perhaps if you have a vegetable plot under attack from a variety of climbing pests, as opposed to flying ones, you might erect a fence especially in order to add some of these spikey things and keep your veggies safe. You might even set up a little ring of them around the edge of the bird table, to see if they'd defeat a squirrel, though I rather doubt that it would succeed for very long.

The same fencing-off idea also comes in electrically powered form as ... well, an electric fence. This can also be mounted on the top of an existing fence or used to create a shocking fence. You'd have to take care to avoid stray bits of foliage coming to rest on the wires, as it will make your garden look very untidy, but it is spoken well of and people also use them to protect garden ponds from herons. And small boys. A slightly cheaper version of this is to use netting, which is probably worth trying before moving on to the more expensive purchases.

More sophistication comes in the form of electronic devices that emit high-pitched noises that you should be unable to hear yourself. These do appear to divide opinion: some people find them to be the miracle cure and some people find they are a waste of space – worth a try, perhaps.

The really fun idea, of course, is using a water pistol. The people who suggest this kind of thing clearly have too much time on their hands, waiting for hours for a cat that doesn't notice you hiding with a loaded weapon. If you need a longer range, a hose might do it, especially if you use one of those hoses with an end that is designed like a pistol and can therefore be aimed before shooting. There are also the sprinklers that shoot water at pests when the spray is activated by an electrically powered motion detector. They are quite popular and are reputed to work against other creatures like deer and foxes.

Perhaps the most entertaining – if cruel – idea to contemplate, but hardly to put into effect, is to insert cocktail sticks all over the garden. The nicest idea is to feed the cat, since they apparently don't poo where they eat, although it wouldn't stop them eating the birds and committing other feline crimes.

ᏋᏫ WHAT TO DO ᏭᏮ

Try simple ideas first: netting, reducing bare soil areas (and therefore toilet sites) and using water sprays

Fasten a plastic bib around its neck to stop it catching birds if it's your own cat. This makes it look rather like a furry traffic cone and the cat will leave home in disgust

Get a dog instead

ᏋᏫ WHAT NOT TO DO ᏭᏮ

Try expensive preventative measures before examining all your options

Use deterrents that could also end up harming the cat

Deer

*I*n Italy you will soon be allowed to shoot deer with a bow and arrow. No gun licence and all that fuss. You'd just need to practice a lot, since your skills are tested before you can go out and play Robin Hood.

But why would you want to shoot a beautiful Bambi creature? If you live in the city, you may not have discovered what's not nice about Bambi. On the other hand, you may have actually bumped into one, because they are, just like foxes, moving in on you.

Forget for a moment the deer that may at this very second be munching your garden and contemplate driving into one at speed on a dark night. When bright lights approach suddenly out of nowhere, it seems that the instinct of a deer is to leap towards the lights. Which is why it will come crashing through your windscreen, probably killing you both. My husband fortunately survived such an encounter, though the animal was killed and his car badly damaged. Deer were estimated a couple of years ago to be involved in 74,000 road accidents a year in the UK.

In addition, deer may be responsible for the dropping numbers of ground-nesting birds. Their chomping away at coppiced and regenerated trees' costs insects their habitats, and by destroying flowering shrubs they lose butterflies' much needed nectar. They may look sweet, but they also threaten other sweet things such as bluebells, dormice and nightingales by devouring the undergrowth in our woods and scrublands. There are now over two million deer roaming around the UK, mostly at night, doing all this damage. All six deer species in Britain are increasing annually: in 2011 fallow were estimated to have grown by 1.8 per cent, muntjac by 8.2 per cent, sika by 5.3 per cent, roe by 2.3 per cent, Chinese water deer by 2 per cent and red deer doing feebly at 0.3 per cent. (I'll admit it – I didn't personally count them and I can't imagine who did.)

As we are principally concerned with gardens here, I will focus on deer's effects on them. Deer are well known for loving roses, just as people do, only people don't usually graze on the roses as they pass by. The blighters can clear a whole rose garden overnight. They are the largest garden pest and certainly can be one of the most ruthless, destroying your most precious plants and vegetables, including beetroot, runner bean, berries, holly, ivy, rhododendrons, geraniums, acorns, chestnuts, apples and even yew, which is poisonous to sheep and cows.

I mentioned their chomping of tree bark, which they do when food is a bit harder to come by, but they can also damage the tree by rubbing their heads against it in order to remove the velvet from their new antlers. This scores and frays the bark, which can fall off to reveal the inner wood and often stunts growth above the damaged area.

They are not so much the graceful, magisterial custodians of the forest often depicted in art and children's cartoons, but crazed destroyers of all things beautiful.

So how to stop the insurgency? Don't present them with the challenge of 'deer-resistant plants' – their tastes vary just as much as ours and will depend on their curiosity and hunger. Other popular ideas include hanging bags or bits of human hair around the place, so if you see people scavenging round your local hairdressers you'll know why, unless they are bald and hoping to stick the clippings on their heads. Lion's dung is supposed to make a deer run a mile – but how much dung and at what spacing will you need to apply it? Has anyone scientifically tested these ideas? Probably not, since both deterrents are illegal under the Control of Pesticides Regulations (amended 1997). True story.

Human urine is reputed to deter deer, yet rain soon puts an end to that. Mothballs, garlic and creosote have been reported as successful by some people. And as unsuccessful by as many other people. Deterrent sprays may work, but require constant reapplication.

We are probably simply avoiding the inevitable, which is that fencing them out is the only reliable way to let you sleep at night. Well, maybe. Given the squirrel, rabbit and deer damage to tree bark in my wood, I gave up trying to establish new trees. The many mature ones now reign supreme over a woodland floor that is gradually getting covered only by self-sown ferns. It's quite a special look and pleasing to see tree trunks so well displayed, but it isn't what we want for the whole of the country's woodland.

So, fencing. Muntjac will go underneath if they can, so it should turn outwards and be slightly buried at the deer side. How high? A deer can jump nearly 1.8 metres (6 ft) from a standing start, and that is the standard recommendation, so you need an enclosure that looks much like a well-fortified prison (you can use wire mesh to enhance this effect). Alternatively, and possibly more effectively, deer are unlikely to jump into something that looks like an enclosure, so construct two parallel 1.2 m (4 ft) high fences built 1.5 m (5 ft) apart. You need to put the whole lot up as quickly as possible to ensure the deer don't get used to and contemptuous of your barriers (thanks to Margaret Roach for this tip). Margaret points out that it should be possible to garden the bit in the middle, which would dramatically reduce the prison effect. I haven't tried this, but it does make sense. And at last here's the use for that dubious list of deer-proof plants – plant them between your fences as a kind of belt and braces approach: a double put-off.

You may try using tree guards to protect saplings from having their bark frayed, browsed or stripped off. You can buy these at agricultural merchants. Remember to remove them when they begin to strangle the tree. They won't help for very long, though.

✺ WHAT TO DO ✺

Buy a contraption that you attach to the end of your hose that detects all moving critters by their motion – cats, deer, rabbits, you – and squirts them. You have to arrange for the pests to be within range of the end of your hose, but it is well worth trying. The idea of squirting them is very satisfying

Construct a fort

Get a dog, which will scare the deer away

✺ WHAT NOT TO DO ✺

Go to Italy and learn to use a bow and arrow

Gun down the deer – you'll probably end up shooting someone else

Bother with so-called deer-resistant plants as a sole measure

Employ 'ultra sonic' devices that are supposed to emit high-pitched sounds to frighten creatures away. The deer will get used to them after a while

Slugs and Snails

One of my earliest memories is of holidays in the Lake District in a cottage on the edge of Ullswater. In order to reach the exciting things, like the lake and the rowing boat (children were allowed to have fun in those days), you had to take a long, rather winding grassy path down the garden. It was, from memory, always covered in hundreds of great big black slimy, glistening slugs. And a fair few snails, too. The only way to navigate them was to look ahead and run as fast as you could, ignoring what you might be treading on. Such big black slugs are not always damaging to garden plants, as they prefer garden debris of various mentionable and unmentionable sorts, but that does not make me feel less queasy at the sight of them. Or the feel of them. I can tell if I have had the merest twitch of a touch of a slug. The notion gives me the horrors; the reality makes me yelp. They are not nice. Similarly, I can't pick snails up and throw them around like some people do. You might just accidentally touch the slimy bit!

But some folks spend a good part of their lives going around picking molluscs up. They are probably Britain's most pervasive pests, and with good reason. The UK is uniquely designed to promote slug and snail infestation, as they are happy souls in cool wet summers and warm wet winters, which you may have noticed, and as I found out in the Lake District. They are not, however, daft enough to simply lie there and die in a hot summer. Slugs will burrow down in the soil until they find the damp layer. They mostly live in the soil rather than on top of it, which may be part of the problem in our gardens. People imagine that they can see how many there are and believe it may be possible to remove them physically or kill them with little traps.

The damaging slugs are the garden slug (small, black, pale side stripe), the field slug (small, grey and flecked) and the keel slug (twice as big, grey with orange stripe). They like to eat most things you like to grow and there are estimated to be around 20,000 of these slugs in the average size garden. They all (being hermaphrodites) lay lots of eggs, which all have long lives.

If you think your garden isn't an average size, try 200 slugs per cubic metre. Still want to go out at night with a torch hunting the ones daft enough to make themselves visible? I heard of someone with very peculiar dedication who apparently removed 27,500 slugs from a small garden without managing to make any noticeable difference to the slug activity in that place.

So perhaps we have to hope that if you like to set traps of stale beer for slugs and snails to drown in, you are not solely expressing your slightly sadistic tendencies in a way that is unlikely (yet) to get you arrested. Going round at night with a pair of scissors in order to cut them in half, as my husband was wont to do before he knew better, can only be called sadistic and I think perhaps the RSPCA is simply slimeist if it doesn't take such activity seriously and stop the people who do this. Perhaps you should also be arrested for pouring salt on a snail or slug – it's deeply unpleasant, and I'm not sure that putting them in boiling water or your freezer, as I have seen recommended, is very kind either. Although I suppose that if you are going to eat them later you may have an excuse.

Killing things is really not nice, sad to say, but I admit I do hold my nose and jump on a snail if one crosses my path, so I am not an innocent either. Gardening brings out the worst in all of us.

A lot of slug-related activity in the garden is just plain daft. Chucking slugs and snails into your neighbour's garden is one of these, as they are quite good at coming back – you have to take snails further than ten yards to ensure they will get lost trying to find their way home. It is surprising that they return home at all, since you'd think that with a house on their backs snails could camp anywhere. If you are severely ecological, you must think twice about moving them anyway, because getting rid of the nuisance will lead to an imbalance in someone else's ecosystem. The effect is to create a slug vacuum that gets advertised immediately on the slug internet, so some slug will be setting off in your direction as soon as you have created a void – someone brilliantly described this as like mopping up water with the tap running.

Understandably, many people think 'we need a predator'. Like the frog, for example. I have never, despite having had many frogs in the garden, seen a frog eat a slug, but I'm assured they do. As do our friendly, if ugly, toads. We have all, I hope, seen and heard thrushes bashing snails to bits on a handy stone. Newts also chomp up slugs, but I suspect in my garden our one rogue fish chomps up the newts … Hedgehogs are good – unless you are effectively fenced against rabbits, in which case they can't get in. Centipedes, glow-worms (we had one once – bit of a rarity, I'd say) and ground beetles are good predators of slugs, too. You can grow your own predators, like ducks and hens. Shrews are supposed to be quite good at eating slugs. You could even turn them into pets and breed them.

You can also buy slug-killing nematodes. These are tiny eelworms infected with nasty slug-killing bacteria, which you can buy in packets and which you add to water and apply by watering can to slug-ridden sites. This is quite an expensive option, though.

Better still, you can make your own nematode soup and

save lots of money. First, catch some slugs – my parents always used scooped-out grapefruit skins turned cut side down to trap them, and they must have got a few since they were a permanent garden decoration. Slugs find the skins a comforting damp shelter and wait there patiently for you to come and cull them. If you want to harvest them for nematode soup you somehow pick them up (long tongs?) and transfer them into a jam jar with a little water and some leaves for them to eat and sit on – you are not attempting to drown your new pets. Keep going until you have a good haul and then decant them into a bucket with a little water and more leafy goodness. Then put a cover over the bucket to keep them safely inside.

Go and gloat every day and give them a good stir. You want them alive and wet. Keep this up for a couple of weeks. Some of the slugs will inevitably already have been infected with the relevant bacteria by their own resident nematodes, and this will create a good soup for diseases and nematodes to breed and multiply in. If you feel extravagant, you can buy some commercial nematodes as a kind of starter kit and add a tablespoonful, just in case all your slugs have been working out, have given up drinking and smoking and are generally miserably fit with no diseases or nematodes. But that probably won't be necessary.

Then you have the fun job of putting this stew through some kind of sieve so you can stick it in a watering can.

Refrain from drinking this lovely liquor or offering it to your friends as a healthy tipple. Use it to water your most precious plants. Save the sieved contents, return to the bucket and keep it all going in order to be economical with your slugs.

You've seen the flaw in all this, though, haven't you? It still creates that ecological vacuum for new slugs and snails to fill. Apparently, the slime is the signal. If your pets have chomped away so successfully that there is a great absence of slime trails, any slug looking for a place to hang his hat will rejoice and move in. Though if there are slime trails, they apparently use them to slide merrily along in order to save their own slime anyway.

Some people recommend getting rid of places where slugs and snails may hide. Under mulch, under bark, in damp places (take out a hair dryer? I imagine you could even dry out a slug with the drier while you are at it), under flower pots and the compost heap (that's another good reason to give that chore up). It may be worth trying to keep your soil dry by watering the plants rather than the bed, supposing your garden is one of those unpleasant ones with bare soil between the plants. You may have to find out how to stop the rain, though.

But this still ignores the basic problem: slugs mostly live underground.

Another approach is to actually try to protect the plants you really care about or which slugs and snails like best. If you still have any. This is quite a popular approach. You can add an electric mollusc fence to your electric deer fence and electric rabbit fence – one at every height, then. I suspect it will be really critical and difficult to keep this one clean of debris, which could inadvertently act as a handy little snail bridge.

You can use boring old non-electric barriers, too. There is a galvanized steel slug fence that looks pretty cool. It has a bent over top that apparently slugs and snails can't navigate as it involves being upside down and then going up over the edge. Just be sure you don't install it the wrong way round and provide a ski jump for them. And having installed it, remove every slimy pest inside – and don't forget the eggs.

Many other barriers are recommended: gel repellents, for example. It is also suggested you might water your plants with coffee, which, while extravagant, is a little more pleasant to brew than slug nematode soup. Or you can get a sticky tape, which may or may not be copper, that slugs and snails are alleged to hate. Copper is an expensive commodity, so it seems strange that such a thing could be affordable. Some people find it works. Some people find it doesn't work. Some people have witnessed slugs slugging right over it, cheerful as ever. Strangely, someone has decided to sell this kind of tape with the edges all cut up to create spikes. You'd think if it worked at all you wouldn't need to do that, though ...

In the Middle Ages, England became so wealthy selling wool that the speaker in the House of Lords still sits on a woolsack in honour of what wool was worth in the fourteenth century. Today, wool has reached such a sad state that people use it in little pellets to scatter around plants to keep slugs off. How the mighty sheep has fallen.

There is also gel, diatomaceous earth (also known as DE or diatom, which doesn't persist in the wet) and many other home-grown remedies that I won't bore you with, since it seems slugs and snails can glide over even glass shards and the edges of razor blades. However, we know something that works with vampires and that may well work with slimy pests: garlic. This can be another home brew in a bucket for you: squish two garlic bulbs into 1 l (1.7 pts) of water and boil for a few minutes. Strain and cool (this is not for scalding slugs with). Add a tablespoon of that to 4.5 l (8 pts) of water and water on to your plants. Bingo – 'slug go'. Best for small gardens.

You may be feeling a bit fed up by now and ready to give up gardening for knitting. However, one of the questions I get asked over and over by visitors to our garden is 'how do you keep the slugs and snails off your hostas?' And the boring answer is – drum roll – slug pellets. I think many people believe that slug pellets are a greater evil than slugs and this may be because they need to be used with care and discrimination. Start by using them very early in the season as a tiny slug hole gets bigger and bigger as the leaf expands, so you want to be quick enough off the mark in spring to avoid that. Or the total annihilation that you may get if the slug or snail reaches a plant before you and while the plant is still tiny enough to be totally devoured. So anticipation is the key. Then, all you need is a thin scatter and repeated application every few weeks. I say thin scatter – it really doesn't take many, and if you scatter well it would take a very determined animal to find and eat up very many.

You may well now be imagining death and destruction, poisoning and misery, all inflicted by those wicked little pellets. Well, they are chemical, so if you try to avoid all chemicals, while excluding the active compounds of garlic, you may be stymied. I find pellets based on ferric phosphate are effective and they are described all over the place as 'organically approved'. However, you may find it quite impossible to discover just who gave them this approval.

As far as I can tell, there are good things and bad things about these pellets. The active ingredient itself, iron phosphate (or ferric phosphate), seems to be reasonably safe, in that it doesn't poison earthworms, and the dying slugs or snails conveniently disappear themselves so that they are not lying around on the soil surface to be eaten by other creatures, like birds. It is described widely as non-toxic to animals and wildlife. It may well be, but it comes packaged with a chelating (don't ask) compound commonly called EDTA and it seems that this is dangerous. It may be toxic (note the 'may' – the definitive evidence is still needed) to earthworms and if to them, then also to good beetles, millipedes and woodlice. And to animals (dogs, for example) or humans if they eat enough. And remember, that might not be in just one go. A dog could be eating it regularly with the same effect. It may even be worse in this regard to the other type of slug baits that use metaldehyde and which are generally regarded as having been superseded by iron phosphate-based pellets. The metaldehyde ones also use stuff called Bitrex, which makes the pellets taste bitter. This has been regarded as unnecessary in iron phosphate-based pellets but could well be a mistake, as there is now some evidence of dog poisoning, which a bitter taste could prevent.

So, if you want to risk either of these baits, bear all that in mind. Keep pellets locked away from children and pets, and use them sparsely. Never pile it in heaps. Remember your neighbours' pets as well as your own. Clean up any spills. Be careful and sensible. And always read the instructions.

✺ WHAT TO DO ✺

Remember slugs and snails are not that bad if you don't grow veggies, hostas or other lush plants

Try slug pellets if you want real results, but beware of using too many

Have fun making horrible smelly potions

✺ WHAT NOT TO DO ✺

Believe there is a quick remedy. Slugs and snails, like taxes, are always with us. Whatever you do to kill them, you will need to continue to do as long as you wish to grow undamaged plants

Moles

I want to be tolerant of moles. After all, they only make little molehills, which I can kick over. That and a long line of subsidence where they tunnel ... But, hey, they are sweet little creatures, so who would want to harm one?

Well, after they had not only disfigured a quarter of the meadow (which is quite big in garden terms) but also totally destroyed all the bulbs that I had planted there, I wanted to harm them very much.

You will have realized by now what a horribly unreformed person I am, so you will not be surprised. But you might wonder if I tried anything else before deciding death was the answer. Well, I did.

I tried a lot of daft things. I tried hosing water down their runs. We have pretty good water pressure, but I suspect the water drained harmlessly away just as fast as I hosed, and in the end all that happened was that I got wet using the hose, as you always do. Well, I do – I never get as wet as when we have a drought and I have to water plants.

I then thought about gassing them. At the time, I had a motorbike and I considered somehow rigging an outlet to the exhaust and pointing it down the mole hole. I probably would have just succeeded in terrifying myself. Anyway, gassing should only be attempted by trained professionals, and, sadly, I wasn't qualified.

I decided to spend money on the problem. I bought several posh-looking mole deterrents in the form of battery-powered devices that sit in the ground, vibrating, and supposedly scare moles away from an area of 1000 square metres. I was overprotected, you might think. Well, the thing about a meadow is that it grows grass and other meadowy things, and any sign of moles and the vibrating metal object that was, I hoped, terrifying them disappeared. The grass grew two or three feet high and I forgot about moleys, having other garden problems on my mind. Until the time came when I had to cut the meadow, which I do with a ride-on mower ...

Yes, you've got there before me. It wasn't going to do the mower much good trying to slice through a solid metal pole sticking up out of the ground, was it? So, where was it?

I ended up having to poke carefully through the whole meadow to find a sticky-up metal object. In the process, I found many sticky-up giant molehills. It was an experiment never to be repeated.

I know that some people have had success with such implements. It seems that no matter what the weird and wonderful (and expensive) thing is that is supposed to harmlessly and miraculously scare pests away, there will always be someone who swears by it. And probably someone a little sceptical but who still thinks it may be working … and the rest who think it's rubbish. You never seem to get everyone thinking it's rubbish. I wonder why that is? The problem can be much more serious than a mole scarer that doesn't scare moles, but people can stay convinced. Iraqi security personnel went on using a phoney bomb detector after they were told they were fake. 'Whether it's magic or scientific, what I care about is it detects bombs,' declared Major General Jehad al-Jabiri. Anyway, if you try the mole scarer, be sure to erect a large flag showing you where you hid it.

You can get granules coated with castor oil as a mole deterrent. I know that small children used to be dosed with castor oil to 'keep them regular' once upon an unhappy time, and perhaps moles are deterred by unhappy childhood memories of their own. But castor oil is dangerous stuff and is dangerous to the workers who harvest it. Is a poison with no antidote best avoided? It is a matter of personal choice. Then again, many people think it is a miracle product for everything from hair oil to inducing labour. Or frightening moles.

The French blow moles up. Seriously. I'm not sure it's worse than mole traps, but it could be much more dramatic if it went wrong. Similarly, a neighbour of ours used to get his shotgun out and shoot when he saw the molehill move. A ricochet from a lurking underground stone could have killed him as well as the mole. Now, I know that pests can be heartbreaking, but perhaps we must reflect at times and confront ourselves. How often do we quietly want rid of something and how often do we really just want to punish the bloody things!

I forget what else I tried before I saw the meadow full of molehills and, unlike the other three quarters of the meadow, not a single hopeful sign of a bulb poking through. This was not just distressing, but expensive, too. I am still not quite sure what that particular mole was up to. I understand they don't hunt in packs, so they must have effective mole-scaring tricks of their own to keep a place to themselves, probably involving lots of tooth and claw. This means that one demented mole had built molehills like mad and done what with all those bulbs? Sold them?

So I found a mole catcher. A real-life authentic country mole catcher. So authentic that he couldn't drive and had to be regularly collected and returned. They don't just visit once, proper mole catchers, they come regularly. Sometimes to set traps, sometimes to inspect traps and sometimes to dispose of dead moles they have caught in their traps. There are no doubts with moles catchers, unlike with deterrents. When they have been successful, you know it and have the proof. And after all that messing around I had done, it was a great relief.

But, I thought, what about when this wonderfully helpful gentleman retires? It seemed as if that might not be too far away. Hesitantly, because it might have seemed like we intended to put a good man out of work, my husband asked if he would teach him his skills. He was delighted to be asked. He was in the sad situation that many skilled men must have been in over the years, of knowing there was no one who would continue his work or benefit from the knowledge he had acquired.

So I now have a resident mole catcher, and a very good one he is too. It seems some people can do it and some can't, and it may have to do with how smelly they are. If you contaminate your traps with your smell so that the mole is warned, and moles have sensitive noses, you will never make a champion mole catcher, so rub your hands and the trap with soil, which should hide your human scent.

❧ WHAT TO DO ☙

Try to find a local mole catcher

Learn a new skill with a mole trap and be ruthless. It works

Level your molehills before mowing

Keep the faith – one of your options is bound to work, even if all else has failed, so continue trying different remedies

❧ WHAT NOT TO DO ☙

Try weasel droppings as a deterrent

Plant caper spurge (Euphorbia latifolia) as a deterrent. Look, even if this works, what is the chance that you'd want to plant it in exactly the spot that the mole had chosen to foul up? And even if it were just the right place, caper spurge is a biennial plant. You know, one of those unpredictable plants that will decide itself whether and where it wants to grow. But you'd be unbelievably lucky to get it growing regularly in just the right place to deter a mole – the same applies to planting garlic all round the edge of your garden

Plant empty bottles or children's windmills in the ground to create an annoying whistling sound for the moles

As the deluded frequently suggest, put mothballs in the trenches. Well, go on, then. Try it and see

Voles and Mice

Voles

I 've never had any trouble with voles. I have encountered them, however, as our cats used to present us with some nice dead ones for our supper. But speaking to friends, I discovered that voles cause a lot of headaches, so they have managed to nibble their way in here.

There are two different kinds of vole that you might find in your garden: the short-tailed vole, which, unsurprisingly, has a shorter appendage than its cousin the bank vole, whose tail is in turn shorter than a mouse's (the short-tailed vole must feel terribly challenged). That, as well as the vole having a rounder head, is how you tell a vole from a mouse when you meet one.

These nocturnal land lovers must not be confused with the water vole, which is endangered and protected (the two terms do inevitably appear together, although we must assume that the endangering comes *before* the protection). If you were to have a water vole in your garden, you may be experiencing some problems, large amounts of running water being one. The other issues will be less to do with the water vole and more to do with its protection, which means you have to leave the banks of your running water rough and uncleared. So no paddling in your little brook or growing nice waterside plants, or doing anything for that matter, even though the voles' burrowing may undermine the banks of your stream. You may end up thinking that the real pests are the intrusive inspectors and officials ensuring that you are not interfering with your water vole, perhaps with the assistance of stern notices to tell you how to behave in the possible presence of said voles.

But the voles that are plaguing my friends are the ones that live on land and eat things we wish they wouldn't. This is where mice come in, because they have a similar diet and, if we are lucky, we may deal with them the same way.

Mice

These are the critters with tails longer than their bodies and come in house mouse, field mouse, wood mouse, harvest mouse, and dormouse varieties. They are often a species of wildlife that people are eager to attract to their gardens, although there are those who invest a great deal of time and effort in hunting them down, too. I don't imagine many people wish to invite house mice into their homes, however dedicated they are to the preservation of wildlife, yet in reality mice don't need inviting, and if you think you can block all available entrances, bear in mind that they can climb bare brick walls and get through a hole so small you can only insert a pencil into it.

Here, though, we are concerned with those that are eating your peas and beans or invading your greenhouse to feast on your newly sown seeds. Field mice and bank voles will eat bulbs, corms and tubers, and there's all the more reason not to grow tulips in the ground outside as these are their special favourites. They will also eat strawberries and other soft fruits, as well as vegetables like brassicas and sprouts. A friend of mine found her climbing beans suffering from sporadic collapse and sudden death, which is when she found they were being chewed away at the bottom of the plant, a common target for mouse attacks.

Both mice and voles will strip the bark from your trees and shrubs, especially in winter when food is scarcer, shinning up them quite happily to get at some good bark above that which the rabbits have already been gnawing at. The risk is that if a stem is gnawed all the way round then any further growth will be killed.

If they are feeling too lazy to climb

they may burrow and eat the roots of your trees and shrubs, just for fun. It's a wonder any desirable thing in a garden survives, so eclectic is their taste. Indeed, mice and voles also eat insects and weed seeds, so they are not always unhelpful. This will no doubt confuse you and undermine your resolution to deal with them properly.

Yet some people smile kindly, pleased to see evidence of happy wildlife in their garden. They see the mice scampering around grabbing the fallen seed and debris from the bird feeders and make a mental note to put some food out especially on the ground, in case they get really hungry.

These types will also opt for prevention of damage rather than major killing sprees, for instance by wrapping their seed trays and pots of bulbs in wire or (green) plastic netting. Or popping a transparent plastic protector over the propagator and making sure the little ventilation cover is not open too wide. Similarly, they will be out with the netting, protecting their trees and winding it round every stem of their shrubs. If their banks are getting undermined they will lay wire netting over the bank to protect it.

These are possibly not the same people who, when they see a sweet little mouse or vole, are thinking how happy the buzzards or the owls will be if they can catch one. Though I believe that like us, buzzards actually prefer someone else to do the killing.

Then there are those who have suffered serious damage

and will be out for revenge disguised as 'control'. This means killing, in case that soppy but ubiquitous euphemism deluded you. Poison in the right places works, but using it anywhere outside risks a nice creature you never really wished to harm, like a small child, eating it and suffering unpleasant consequences. Also, between spring and autumn there are other food sources available to the mice and voles, which might make poison less tempting.

Therefore trapping may be better. You can usually successfully execute a creature of this size when you trap it, although the whole thing is mightily unpleasant. The danger with traps in the open is that they will catch a bird instead. They are also prone to getting wet and rusty. If you are serious about this, it's best to make a box to hide your traps in. You can cut a small mouse-sized hole in a plastic lunch-box container and place your trap and bait in that. It works.

Or you can build yourself a wooden one with a top you can handily lift off, and which can take several traps. This has the advantage of smelling mousy or voley, which is said to be a better attractant than the usual food bait, which they can become more cautious about. I've found peanut butter does a good job and a number of times I have, oddly, found the eaten bait alongside the dead mouse that presumably ate it. This suggests they at least died happy. It seems all of God's children love peanut butter.

᭡ WHAT TO DO ᰰ

Try apples, carrots or oats to catch voles, which are harder to trap than mice

Stay cheerful. Mice are curious creatures and after many, many generations of their ancestors getting caught in traps, they still manage to get caught the same way

Use physical protection where possible, like covers over seed trays and spiral tree guards on young trees

᭡ WHAT NOT TO DO ᰰ

Keep mice and voles off your plants with rigorous garden hygiene. This means offering them no cover of mulch, fallen leaves, general debris or decaying plants. This is a technique called 'no place to hide' and is not very garden-like. Don't forget to also polish your seed trays and dust your spade

Get too attached to the mice in your garden, although it is perfectly acceptable to attempt to prevent rather than destroy if you don't object to a lot of extra work

Forget to empty the traps every day. They get smelly

... AND SMALL

Biting insects ... and Snakes

Midges

A lot of people avoid Scotland simply for fear of being bitten. Not by a Scot, but by midges, which apparently even come with a special spelling up there: midgies. The women midgies notoriously gather in swarms and bite people, as well as cattle, sheep and anything else with nice warm blood.

They have the peculiar characteristic of making smokers popular, as they are as oversensitive to cigarette smoke as an overzealous government official. Regulations emanating from such officials also mean that one of the effective deterrents is one of those open secrets, sold as 'Skin So Soft' to the great embarrassment of macho Scottish farmers, who are reputed to keep the pretty bottle in a discrete corner of their quad bikes. It can't, being a skin softener, be advertised or promoted as an insect repellent, hence the nudge-nudge aspect of using or speaking about it that way. There is some talk that it may have become less effective lately anyway, but like all these things it works for some people and not for others.

It's said that midges are attracted to dark clothing, possibly to HRT, gloomy, wet places and carbon dioxide. The cure, then, is to stop breathing and wear a white shroud. And bog myrtle, I am told, despite a certain reputation, is completely useless, although wearing it may look jolly.

The more conventional solution is DEET, which is claimed to be very effective, even though some people don't like the smell and the greasy feel, or maybe just slathering themselves with potent chemicals. It is also reputed to melt plastic. An alternative, which is given respectable support, is a repellent based on lemon eucalyptus. The very popular citronella is not always effective, but it is based on lemon eucalyptus, so work that one out.

You can obtain a thing called a Midge Magnet, which attracts midges with carbon dioxide and then vacuums them up. As it uses propane gas to produce the CO_2, you will have to cart a canister and machine around with you. So maybe this option won't be the one for you.

Horse flies

*I*n most of the UK, horse flies are possibly the worst flying menace – and they really do have a habit of coming up behind you to rip you open. Yes, they don't just give you a little bite; they slice you. And potentially give you diseases as well as the most horrible huge itchy lump for days.

The pleasure associated with horse flies is that they are a little slow to start, and if you're very quick when one lands on you, you can slap it (and yourself) hard and kill it. Which is incredibly satisfying.

The weirdest solution to these revolting things that I have heard involves a bright blue flowerpot. It seems that horseflies (also known as deer flies, as they like vampiring deer, too) are visually attracted to moving objects, and especially bright blue, perhaps because it rather stands out amid the variety of greens in the places (damp, dank, gardeny) horseflies hang out. You smear the flowerpot (size unspecified) with something nicely sticky – maybe Vaseline, tree grease or auto grease – and then stick the pot on a pole and go hunting. Horse flies reputedly will flock to your pot and stick with it.

For those too encumbered to parade a pot you can also try a bright blue plastic cup (there's a nice thing to search for – in picnic sets?) attached to a baseball cap. I have not tried this. I imagine, though, that it's a good ice-breaker at a dull party.

Bees and Wasps

What on earth do you do when a flying sting starts buzzing round you and everyone tells you not to move? Sit still and make a perfect target? Dreadful, impossible advice. Better get up and run away as fast as you can.

I think all advice about how not get to get stung by wasps and bees is depressing: don't smell nice, don't wear colourful clothes, don't have cream teas out in the sunshine, stay out of flower gardens, don't try to hit them and sit still. And all of this is very serious for those people with allergic reactions – which apparently get worse with every sting, instead of inoculating you against the poison.

I have no bright and useful suggestions here, except to follow all the good advice. I am the idiot who trod in a wasps' nest twice. There was a tantalizing weed about a metre into a border, so I stepped in and reached out. My foot went into an underground wasps' nest and many wasps emerged, determined to exact revenge. I ended up in A&E. You'd think that would be a good lesson. But next week that weed was still there, taunting me. Unthinking, I stepped into the border – and you've guessed the rest. I am clearly not a person to take wasp advice from. Except that any unusual or excess reaction to any sting merits treating as an emergency. People have died of stings and bites. So, panic.

Otherwise, apart from stinging people, wasps and bees are a Very Good Thing. People are gradually recognizing this and are using plants bees like in ever increasing quantities. But wasps are also pollinators, and just because they get bad tempered at the end of the summer, like the rest of us, that is no reason to destroy them unnecessarily. You may well call in someone to exterminate a wasp nest too near your house, but elsewhere, think twice. You don't have to poke your foot in that nest in the border, after all.

There is also good news for bees, it seems. An insecticide based on spider venom and, err … snowdrops (so I believe) has been developed and we are assured it is harmless to bees. Well, I know, we've been here before and heard many such things. But as long as farmers use pesticides, it is great news if there are possibly some harmless ones appearing. So keep your fingers crossed. And are some people more likely to get stung than others? Well, apparently people with smelly feet are.

Snakes

All right, so snakes aren't that small, and aren't insects, but they do have a nasty bite. Adders are the only venomous snake you might meet, and people don't often die of their bite. Still, best not to jump up and down on one if you see it. Admire from afar.

But, if you do get bitten, remember that nasty alien snakes sometimes do manage to wangle their way into the UK.

So, a snake bite? I'd panic again. I'm not sure what the big thing about not panicking is about. Sometimes when you read the government's thinking about various eventualities, it appears that panicking is worse than a bomb or a flood, which may precipitate a good panic. I suppose it may stop you thinking clearly, but on the other hand it may motivate you to act when it is really necessary. And the British are not good at that.

I remember sitting in a London pub during the IRA bombing campaign when a voice came over a loudspeaker urgently saying that if someone didn't move that suitcase quickly, it would be removed and the police called immediately. A slight cause for panic, you may have concluded. No one moved; all went on as before. Better to be blown up than be the one who just got up and walked out to safety.

❧ WHAT TO DO ☙

Panic and never step in the same wasps' nest twice

Experiment with deterrents until you find one that works for you

Slap horse flies (not bees) fast

❧ WHAT NOT TO DO ☙

Tread on a snake or a slow-worm. They're not that scary

Wear shorts to trim your hedges near where you think there are horse flies. You will regret it

Ignore a rapidly swelling bite or sting – especially one anywhere near your mouth or throat

Aphids

I'm squeamish. That makes dealing with garden pests a special challenge. You will come across people who take great pleasure in squishing things in their fingers in front of you. For some of us, this is a traumatic experience, and not something that will make for a pleasant stroll around the garden. Some of us will run away if you squish things in front of us, although there is a chance we might also be grateful and appoint you Chief Squasher.

So you won't find me squashing an aphid (also known as a greenfly) with my fingers. But years ago, I found a way round this if you have (like me) relatively short arms and you keep your sleeves long. When you see that green mass seething up a stem, or notice the aphids have mottled, yellowed, wilted or stunted a plant, pull your sleeves over your hand, grasp the stem below the obscene wrigglers and squash them all with your flesh protected by the sleeve, which goes green. You may have to wash the greened garment subsequently, so it might be worth having a good hunt for more of the bugs before you go indoors to make best use of the one wash, as it were.

Of course, you could wear rubber gloves for this exercise. You can keep your hands dry and safe from slimy things or nettles hiding unexpectedly in a clump of innocent-looking foliage, if not from fatal thorns. However, after five minutes your hands will be sweating and you will be longing to rip off the rubber to give them a little air.

There is also considerable delight in employing a hose if there is one handy. A finger over the end controls the drama of the spray, which can then be employed to scour all the greenfly off your plant. The power wash effect may damage the plant, however, and if they are lupin greenfly, they will simply heave a resigned sigh and crawl back up to their feast. But it is immensely satisfying in the meantime.

You will be delighted to know (especially if you enjoy a squish) that there are 4,500 species of aphids in the world. Just why they decided to stop proliferating the variety of their species at such a neat round number is one of the mysteries of life, but we must be grateful to the aphid counters for the information. Aphids also transmit viruses to your precious plants, just in case you thought you'd ignore them and leave them for the birds. (Though, truth to tell, I usually do just that, as I'm not particularly fussy about the odd bit of damage.)

Ladybirds are reputed not only to eat greenfly, but also to farm them, so watch out for tiny tractors zooming

up and down the stems of your plants. They lay their larvae among the greenfly; the larvae then hatch and have a feast. I have, with this in mind, endeavoured on several occasions to introduce a ladybird to aphids on my plants with absolutely no success. I carefully catch one on my finger, having convinced it that I am a plant, and then tiptoe to the target where I have to try to persuade it to let go and drop neatly onto the right leaf, which has temptingly fat aphids waiting in docile ignorance of their planned fate. However, the ladybirds appear to prefer just to drop off and vanish, despite my kind efforts to provide a flock. Maybe they are fussy about the aphids they will farm, as farmers interested in particular breeds of cattle or sheep may be. Using the ladybird as a predator is probably something you'll have to leave up to nature.

And, in fact, it does appear that ladybirds are not as good at harvesting aphid eggs as they are reputed to be, as aphids out-lay the ladybirds by about three to one. To be effective at eating enough aphids, you need to bring in your ladybird very early, before their prey have got a grip. And probably before you noticed there were any.

Frogs like to eat ladybirds, which rather demonstrates how difficult a balance in nature can be to achieve, as everything will have its natural predator. The ladybirds themselves can become a pest. In hot, dry summers they can reach plague proportions and people have discovered, often for the first time, that ladybirds can bite. Presumably, having eaten up every available aphid, they decided people were the next best thing.

Worse, there are also poisonous ladybirds. The harlequin will not only chomp you, but also our sweet aphid-farming ones as well as butterflies and moths. Their chomp is a minor and irritating stinging sensation, but, inevitably, some people are allergic to such an attack. So beware and cheer on the swifts and swallows that eat them with impunity.

The other creatures that farm aphids (they are clearly the plankton of the insect world) are ants, which, unlike ladybirds, do them no harm. Ladybirds are a bit like the factory farmers of the insect world, intent only on devouring, whereas ants take care of them for the sake of that which is gracefully called 'honeydew', but is, rather anti-climatically, greenfly excrement. It is, apparently, sweet, hence the name, and some ants are so keen on it that they 'milk' the aphids, stroking them with their ant antennae to get them to defecate. It even seems that some aphids can no longer poop without such assistance.

So what to do about your greenfly invasion? If they are devouring something very precious, try the squish-regularly-and-frequently method, or in extreme cases spray with an insecticide or water containing some washing up liquid, bearing in mind previous warnings.

But if they are outdoors and you can risk the plant, leave them for the birds. Or cut off the infested bit of plant and stomp on it, for your own satisfaction.

If you get the lupin aphid, on lupins or one of the shrub lupins, get rid of the plant. Those monsters will surely beat you otherwise, as the physical controls used on regular aphids may not work. They are huge for an aphid – up to 4 mm – and you will know they are there when your plant turns into a seething mass of obscene insects and begins to wilt. Interestingly, ladybirds do not seem to be as tempted by the lupin aphid, meaning chemical control is your best bet. When you see them relaxing on the underside of the plant's leaves from springtime, spray them with thiacloprid or acetamiprid, or organic sprays comprised of fatty acids or plant oils. It is best to avoid using chemicals when the lupins are flowering as this could harm the pollinating insects.

⤞ WHAT TO DO ⤝

Ignore the greenfly. Birds eat them and feed them to their fledglings, so you are doing them a service

Squish or spray if they are a real pest

If the aphids, or similar, are destroying a crop you can use a physical barrier in the form of horticultural fleece once you've squashed the lot

⤞ WHAT NOT TO DO ⤝

Wash the plant in a bucket of soapy water. As long as it's not a rose bush, is it going to be easy to turn upside down into a bucket?

Plant lavender underneath (presumably very quickly), which certain 'experts' tell us keeps greenfly away

Buy some ladybirds

Let the cat sit on the horticultural fleece

Beetles

Flea beetles and Carrot flies

I happen to be fond of rocket, so there was a time when the idea of being able to casually wander outside to pick some for a salad seemed a great idea. That was before I discovered flea beetles, so named because they are the gymnasts of the vegetable world. They eat chunks out of your rocket leaves, leaving them looking like they've been peppered by machine gun fire and making them pretty useless for human consumption. And when you go to pick your holey leaves the beetles will advertise their presence by leaping off in a spectacular fashion, preventing you from administering a lethal squish. Even if you aren't inclined to be organic you are not going to want to eat leaves chemically sprayed to kill the beetles and if you are organic you'll certainly find them a major challenge. There are varieties that consume other brassicas (cabbage, sprouts, radishes, broccoli), stocks, wallflowers and no doubt much else, and they also generously gift your plants with bacterial and viral infections.

Plants are most vulnerable to the beetles when seedlings, so, if possible, ensure germination and development of the plant as quickly as you can in order to get through the trickiest stages rapidly. If you don't care what your veg plot looks like, you can cover your plants with fleece supported by hoops, or some form of insect-proof netting. This works against carrot fly, too. You need to do it in anticipation of an insurgency by the pests or you may end up simply providing them with a cosy tent to live under, along with their larder. Try to keep your plants watered during dry spells (obvious, I know) and ensure you don't sow into cold soil.

There is one method often recommended that takes advantage of their leaping habit, which is to smear a piece of board with Vaseline, hold it sticky side down over the plant and then nudge the plant to set them leaping on to the Vaseline.

Alas, if you have left it too late and they are there, jumping up and down, or in the case of the carrot fly, producing grubs that will munch your carrots, you are often advised to try psychological warfare. Confuse them. Female carrot flies are attracted by the smell of surplus carrots being removed, so the advice to sow your seeds sparsely to avoid thinning the seedlings may itself seem confusing as it might sound as if you will yield fewer carrots, but it will mean you won't have the boring task of thinning and you won't alert the scent-sensitive carrot fly to your baby carrots. You should also try to rotate your crops so that flies wintering in the soil don't eat into and grow within the carrots.

So, if your carrots (or similar vegetables – parsnips, parsley, celery) reveal brown cracks or scars in their roots and when you open them up you find tunnels filled with long, thin maggots, it's time to take action before the whole lot rots from the inside out.

Lily beetle

Lily beetles are a lovely bright red colour, but don't be fooled. Together with their offspring, they are capable of stripping your lilies of foliage, which is less lovely. That equally relates to plants like fritillaries and martagon lilies (I know you know they are lilies, but you may have forgotten). The beetles have revolting babies, which are very poorly brought up and never get beyond the stage of hiding themselves under a coating of their own black excrement. (Someone should be urgently discussing this on Mumsnet before this bad behaviour spreads.) The grubs live on the underside of the leaves, which they will eat from tip to stem, whereas adult lily beetles make round holes and will feed on the petals, too.

The scarlet horrors are easy enough to see, but catching them is a different matter. The slightest hint that you are about to grab and squash them – squashing being the recommended kill method if the infestation isn't large – and the beetle will simply drop off the plant. You'll have a job finding it down there unless this lily is in a pot with a nice clean surface of compost for lily beetles to drop on to. In a border you have no chance.

The trick is to go prepared with something you can easily slip underneath the red devil, ready to catch it. Then try and squish, if you like, or just poke and catch it with your hand underneath. Then you can also tread hard on it, which is much easier than crushing with your fingers, especially if you are squeamish. You have to hunt every day, though. The excitement of the chase doesn't last for long and you soon begin to question if a lily is actually worth it.

For more serious infestations, it may be better to use an insecticide, as picking so many beetles off the plant by hand becomes unfeasible, not to mention tiresome. As with all plants, avoid spraying them if they are in flower so you don't inadvertently harm a friendly bee.

Asparagus beetle

These come in lots of pretty colours – red, cream, dark blue, even patterned, and have revolting caterpillar-like children. They eat both the leaves and the bark, stunting growth and turning the leaves above brown. These and the offspring can all be searched for and squashed as with the lily beetle. I only discovered these pests when I grew some asparagus and my spears got all curly and inedible. I gave up in the end, but if you prefer to keep growing it, then consider burning old stems at the end of the year to avoid beetles spending their winter on your veggies.

Raspberry beetles

These are those white wiggly creatures that appear to say hello just as you are putting your spoon to your mouth. They can also affect blackberries and other cane fruit by feeding at the stalk end, mainly damaging summer-fruiting berries. Grow autumn-fruiting raspberries – they don't get beetled as much as summer-fruiting ones.

ᘒ WHAT TO DO ᘓ

Use insect-proof netting or a fleece to hide your crops from flea beetle and carrot fly

Restrict egg-laying opportunities by sowing your seeds sparsely

Squish lily and asparagus beetles even if you're squeamish and they are quick – it's all about technique

Make sure you check the underside of the leaves to find the grubs

ᘒ WHAT NOT TO DO ᘓ

Crush any old beetle. There are lots of good ones, too. Check their colour just in case

Use insecticide on plants that are in flower as you would risk harming bees

Mealy bugs and Scale insects

Mealy bugs

*T*hese are aggressive little bugs that look like little woodlice wearing white furry jackets. Unfortunately, unlike woodlice, mealy bugs drain the life out of your indoor plants and brassicas by sucking up sap as they feed and secreting a honeydew-like substance in the form of a powdery wax layer. Honeydew sounds quite nice – you could imagine feasting in paradise on honeydew, as in 'Kubla Khan' – but we know that this is in fact insect poo. On your plants it then becomes a black sticky business as sooty mould (fungus, of course) grows on it.

All this is as horrible as the pest itself. Watch out for ants, which sometimes share a symbiotic relationship with the mealy bugs. You may find the ants having a picnic on the honeydew, which should alert you to trouble (see Aphids, page 69). Another sign will be that

the leaves of younger plants sometimes become distorted or acquire a whitish-yellow discolouration.

The mealy bugs are tender little insects, though, and they don't like to get too cold – hence their little furry jackets? This leads to our first option for eradicating them: frighten them with a freezing. You do this by putting the infested plant in the coldest place you dare, like an open window in the middle of a big freeze, and then watch the horrors migrate to the place farthest away from the cold and huddle around the edge of a leaf, shivering. Then, in theory, it is a simple matter to harvest the leaves that they are huddling on and burn them. If you're not feeling quite so vengeful, you can simply wipe them off the leaves they have congregated on.

Alternatively, you can squirt them with some washing-up liquid in water, which has the added benefit of giving you the illusion that you are shooting them, as well as causing them to drop off your plant. It may be a good idea to have a bag or sheet underneath with which to catch them, but the logistics of getting it suitably and strategically placed under a plant in a pot suggest that this might be rather frustrating, messy and difficult. They are usually to be found on the underside of the leaves, as well as the growing points, so concentrate your fire here if you choose to employ this method of control.

I get some bizarre pleasure from using a cotton bud soaked in methylated spirits to stroke the little bugs. This usually kills them and feels suitably punitive, as does the widely recommended tactic of hosing them off with water. However, I do think that this is one of those remedies that may cause more damage than it cures – if you're using an effective force of water to remove the bugs, you'll probably defoliate the plant at the same time. And the plant will fall over and spill compost everywhere too, I bet. A safer option may be to wash your plants in the shower, with the water being better contained than from the end of a hose.

If you are feeling particularly cruel, you can try our old friend diatomaceous earth. Some people say this dehydrates them, others that it is like making your favourite pests walk through broken glass, lacerating them to pieces. They will try to get it off their bodies and end up ingesting it and lacerating their insides too. (You can, apparently, put this in water, drink it and rid yourself of intestinal worms while you're at it, although don't do so without, as they say, seeking appropriate medical advice.)

The foes to lacerate are multitudinous – bed bugs, all creepy and crawly insects, fleas, ants, aphids, houseflies. You can add it to potting compost to lacerate anything that might presume to take up residence there; it really is a versatile weapon. It comes in puffer packs and you can puff it over your mealy bugged plants. You can turn the universe powdery white and kill off everything little and troublesome, if you really like. The product is seen as good 'natural' stuff, looks like crumbly rock, and is even used in toothpaste, so you have likely come into contact with it already.

Or you could buy a dedicated insecticide and try that – though the insect's natural defence of a waxy coat means it has to be persuaded to swallow some, so it had better be systemic.

The most hassle-free method, however, may be simply to get rid of the brassicas so that they don't infect next season's plants.

Scale insects

Scale insects offer the usual symptoms under glass – in other words, a miserable looking plant. While there are many different species of scale insect, they all like to suck the sap of houseplants, greenhouse plants, fruit plants and ornamental plants. You name it, they like it. When you inspect the plant, you will see hard lumps stuck to the stems and undersides of the leaves, which are the tough coverings of the insects. Like the mealy bug, they are insecticide proof, as they've seen that coming and developed a hard coat, and can weaken plants and excrete honeydew, which you'll remember helps black sooty mould grow. But you can scrape them off with your thumbnail or – for the more squeamish – a knife blade. Scrape, scrape, scrape every one that you can see and you may have done it. I really believe I have got on top of them that way in the past, but then I may have caught them early. Very satisfying.

You can attack them indoors with a parasitic wasp, *Metaphycus helvolus*, so this might be worth looking in to. Or just scrape, scrape, scrape every day for a while. If they are on outdoor plants, cut off affected foliage and incinerate them (taking care not to let the RSPCA know, in case scale insects are the subject of their latest campaign).

❧ WHAT TO DO ☙

Try meths on a cotton bud if you are feeling patient. Or a systemic insecticide if you can contemplate that with equanimity. Or both

Remember mealy bugs are unlikely to actually kill your plants

Investigate biological control if you'd like some new pets to look after

❧ WHAT NOT TO DO ☙

Get water all over yourself, the furniture and the floor trying to hose mealy bugs or scale insects off the plants. If you try that with a pot plant outside, the plant will fall over and you will also be squirting potting compost all over the garden

Codling moth

When I first grew apples I lived in fear of the codling moth. It is thanks to them that you may find a little white wriggly thing with a fetching brown head emerging to greet you when you bite an apple. So I bought little green triangular plastic houses with a sticky sheet inside and a pheromone bit that smells of the female moth, which you hang in the trees and which bring the male moth zooming to a sticky end.

These were horrid. No matter how hard you tried, you always got some of the sticky stuff on you as you attempted to get it into the little hanging house and then placed the pheromone pellet on it. There doesn't seem to be a solvent known to man or woman that will remove the stickiness from your hands.

So, you hang these eyesores on your trees in late spring and the codling moth arrives to get itself permanently and fatally stuck to the sticky trap, lured in by the pellets, which emit a smell similar to a female moth. Oh, yuk. But what if a small bird should fancy the moth? Might it not get caught on there with it? Fortunately that never happened with my traps, but it did have an amazingly long-lasting effect because I was able to grow lazy and stopped bothering with sticky traps or checking them, forgetting about the maggot-infested apple possibility until now. Which will probably mean I'll get a plague of maggots in my apples this year.

You can also make a version of your very own trap, where green plastic doesn't need to come into the picture at all – although, modern DIY being what it is, another sort of plastic inevitably does. Mix 300 ml (½ pint) of vinegar (any old sort) with 300 g (a generous ½ lb) of sugar and add 3 l (5 pts) of water. Shake and then dispense into plastic (what else?) drink bottles with a 5 cm (2 in) diameter hole above the water line and hang in the trees. The moths dive in and drown in their enthusiasm for sugary vinegar water. It is a cheap enough experiment and any hungry birds would have to hang about outside in desperate frustration, seeing moths but unable to acquire them, unless they somehow squeezed in and joined them. Hopefully harmless moths that would otherwise be on our lists of endangered species don't also come to the party and unwittingly drown themselves.

What else do you need to know about codling moths? That they'll also be after your peaches, pears, quinces and walnuts? That you can also use an insecticide to spray against them, although this will only work on trees small enough to be sprayed and when the larvae have yet to discover that they may like to munch their way into your fruit. You will need two doses of deltamethrin or lambda-cyhalothrin – the timing can be judged by the arrival of the right moth in your trap. The right moth will be just over 1.5 cm (½ in) long with grey wings ornamented with a brown stripe and wing tip. Easy!

⤜ WHAT TO DO ⤏

Cut your apple in half before you eat it and you'll see the horrible worm before you bite on it

Look out for their exit holes as this is a telltale sign of an infestation (or a stray bird having a peck)

Employ more than one method of control – pheromone traps alone might not be enough, depending on how many trees you have

Hang cardboard (any old bits) in your trees in winter for the larvae to live in. Until you burn it (the cardboard) in the spring – if you remember, and can stand the hassle

⤜ WHAT NOT TO DO ⤏

Check your trees regularly and get rid of any infected ones to keep the numbers of nasty things down

Take off all loose bark and vacuum underneath the trees. Then get down on your hands and knees with a dustpan and brush … well, you get the idea

Red spider mite

*I*f you find cobwebs on your greenhouse or precious conservatory plants, don't think kindly thoughts about spiders generously weaving webs to catch houseflies and keep them off your sandwiches. Closer inspection will tell you that disaster has arrived in the form of the dreaded red spider mite. Try a spray of clean water in one of those plastic hand sprays and you will probably find the whole plant is covered with web, suddenly and horrifyingly visible.

If you have a magnifying glass – an essential tool for those gardeners who like to be well and truly scared by whatever crawls and flies around their domain – examine the leaf, especially where the leaf meets the stem. You may see what looks like a little red spidery thing, which is why it's been given such a thoughtful and imaginative name. It has eight legs, I understand, but unfortunately does not come in other collectible colours. The mite likes it warm and dry, just as we do, hence why the greenhouse is its usual abode, although they like to hide in cracks and crevices in walls in winter.

I have fought this creature for years. I started by simply trying to wash it off, initially with plain water and then with the addition of washing-up liquid, which is always worth a try in the first instance. All those webs look as if a fierce spray of the hosepipe would readily displace them, and it's very satisfying to imagine the devastation you are causing in red spider land, but sadly this is not the case. They clearly put on their swimming costumes and have a revitalizing shower, cleaning themselves vigorously if you happen to have also sprayed with washing-up liquid. This is sometimes suggested as an effective combat method if you keep at it every other day. However, it seems they are merely grateful for the quality hotel service, as it never seems to encourage them to pack their bags.

I once tried plunging the plants in their pots into the garden pond, only to find that compost floats. I then had to spend hours trying to fish the resulting muck off the surface of the pond, feeling an idiot, and also finding no sign that the mites had been affected at all. I think you could leave a plant under water for a week and the bugs would just take the opportunity to learn to swim.

Insecticides are often cited as a possible cure. However, I have experimented with all the insecticides I could possibly find, including several that a friend of mine retrieved from the darkest corner of their shed. It didn't make any difference: they become immune to any effective insecticide, should you ever find one, so don't prematurely rejoice if you do manage to zap them with a poison. It's useful to know, too, that some insecticides only kill them at particular points in their life cycle. So the escapees may cheerfully pop up just when you thought you'd got them.

There is also the option of what was biological control, which was often recommended. Perhaps the pest eaters have now been discovered by eaters of pest eaters, as you hear less about this combat method now (see 'Whitefly', page 106, for other uses of pest eaters). On this occasion, I discovered, after some correspondence with pest eater providers, that I may have a very unusual monster red spider mite, indigestible to all known pest eaters. (Lucky me. You now know why we call certain glasshouses 'conservatories'? They are conserving precious and unusual pests.)

I examined the plants again. And truly, I did not need a magnifying glass, although I am short sighted, which always makes seeing things close up somewhat easier, so I could see my precious charges quite clearly. They didn't try to hide, so unafraid of me they were – I'm sure one of them used two of its eight legs to give me a 'V' sign.

Assuming you have the common red spider mite, it is worth at some point trying a little biological warfare. In such a case, you should purchase a nasty little thing called *Phytoseiulus persimilis* – the red spider mite's natural predator – which will eat up the eggs and some

version of the hatched mite for breakfast. Ominously, you are told it needs to be kept warm – over 21°C, which is happy room temperature for us mortals, too, and quite unaffordable in a conservatory except in summer – or it won't eat enough or reproduce quicker than its spidery breakfast. After all, one spider mite can produce 13,000 babies a month, and each egg becomes a breeding adult within eight days at that temperature. The dream is that *Phytoseiulus* should feed until it's eaten every little mite, and then they, well, starve to death.

The fact that that there is also discussion of what insecticides may be used with *Phytoseiulus* has to make your heart sink (some suggest plant oils or extracts, as most chemicals will also kill the predator). But this method is certainly worth a try, if only because it is such a pleasant fantasy that your pest is being eaten by a voracious predator.

In my crusade against the spider mites, I had the bright idea that they might find a steam bath a bit much. I bought a steam cleaner – persuaded by the notion that I would then be able to clean the house with absolutely no effort, as advertised. It didn't remove the glued-on grease from my cooker, but I still hoped that it might frighten the mites. After filling it up with water and heating it to steaming, I dragged it into the conservatory and made sure I got a good angle on those critical meeting places of leaf and stem where the mites appeared to hide. I wondered if my plants would mind being steamed, but reasoned that they would thank me if it rid them of sap-sucking insects. It was quite a satisfying exercise, spraying visible ultra hot steam everywhere I could imagine the nasties lurking, but the results were not instant. There were some days when I was able to go and examine the plants and imagine an absence of lurking spiders. Long enough, indeed, for me to repeat the treatment, thinking that the reappearing pests were simply ones I had missed. No chance – they loved it. Rather than exterminating their population, they appeared to be thriving – after all, what's not to love about a sauna?

So what happened in my battle? I discovered succulents, which appeared (and still appear) to be immune to the mites. Sometimes it is better not to make things harder for yourself and know when to avoid a long slog. I do, now and then, try and reintroduce a flowering plant and discover that the bugs are still somehow surviving. So keeping flowering plants under glass is for me a short-term thing. Bulbs are short-term by nature anyway, wintering elsewhere then flowering fairly briefly, so I do persevere with tulips. I freed myself from the notion that conservatory plants have to be tender some time ago, so various cheerful additions make their way in and out. Keeping them mobile is perhaps the trick.

And garden versions of the mites? Well, why not go and look? They can infest roses, daylilies, chaenomeles, apples, blackberries, box, juniper and – as we have noticed – 'any plant grown indoors'. And, in case you think you might get vicious with them and do some vengeful squashing, bear in mind they stain fabrics and your wall decorations, too.

ᢙ **WHAT TO DO** ᢙ

Grow succulents

Try the red spider mite's natural predator, PHYTOSEIULUS
PERSIMILIS, *if you can afford to keep it warm*

ᢙ **WHAT NOT TO DO** ᢙ

Get the larger variety of spider mite

Bother with insecticides

*Try (as sometimes advised) to keep your greenhouse or
conservatory spotlessly clean – this way madness lies*

*Keep such a distance between your plants that the spider
mites can't migrate from one to another (you cannot be
serious ...)*

Mullein moth

*I*t's quite amazing that there are plants that have their own dedicated pests, but the names of bugs are not always a clue to their favourite food. Earwigs don't eat ears or indeed, wigs. Nor do they, despite the reputation they once had, climb into your ears when you are asleep. Or at least not so often that it gets recorded, and I imagine that if they do, it will be purely by accident. Fireblight (see page 115) doesn't cause fires, nor honey fungus (see page 148) grow on honey.

However, there are still a great many specialist pests, far more than I have ever encountered and far more than I am going to bore you with. Here's some veggie nasties: bean seed fly, blackleg, blossom end rot, cabbage caterpillar, cabbage root fly, carrot fly, celery fly, leek moth, mangold fly, onion fly, pea and bean weevil (versatile, that one), pea moth, pea thrips, potato blight, rosemary beetle, common smut of corn (couldn't resist that one), tomato moth … Puts you off the allotment a bit, doesn't it?

All the varieties of moths make you wonder, though.

Should we really be trying to exterminate them? Moths don't get quite the same press as our sweet furry bees or the often more spectacular butterflies, but they are in rapid decline. Dilemma?

As an ex-veggie grower, I haven't encountered many of the moths with vegetable and fruit tastes. I did come across the mullein moth, though, which has quite a spectacular appetite. Mulleins are verbascums, quite a desirable plant, until the munching starts. The black-dotted, white and yellow-striped caterpillars of the mullein moth are so voracious that they can completely strip a plant, and have such a preference for verbascums that they are being considered as a way of reducing the unwanted wild population of verbascums in America. So the question of how pestilent this creature is should be regarded as quite loaded.

So you may consider whether you really want to grow verbascum and watch it chomped to bits. Or you may think it would be a worthy and useful activity to grow some as a sacrificial plant for the benefit of this particular moth. Which may then, despite it's clear preference, start on your buddleia and figwort after it's finished off your verbascums.

❧ WHAT TO DO ❧

Tempt birds over, as they will eat the mullein moth

❧ WHAT NOT TO DO ❧

Forget that two-thirds of Britain's 337 species of common larger moths have experienced a substantial decline over the past four decades. That's you and your veg plot that caused that

Use insecticide unless you're sure you want to help the moth decline

Vine weevil

I dread that terrible moment when you see a plant looking a little wilted and unhappy and find you can lift it clean out of the soil without hindrance. It has no roots. They have been devoured by the horrible white maggoty grubs of the dreaded vine weevil. You may spot the signs before this if you're clever and vigilant: adult vine weevils eat notches in the leaves of your plants in spring and summer, so keep a look out.

Go and find a picture of a vine weevil and memorize it. Then you will be able to squish it when you spot one. A well-known garden writer once visited us and totally failed to raise any enthusiasm for my garden. He got very excited at spotting and squashing a vine weevil, though.

There was a time when I lived in terror of them – perhaps this was influenced by his triumphant squash. I went hunting them in the woods by torchlight, which seems totally mad now but must have been fun at the time. I think I had a bucket of water and persuaded them with a little poking to drop off the holly they were devouring into the bucket. It was a bit sploshy doing all that in the dark in the wood, and they probably weren't doing much harm up there. The weevils seem to like pot plants best and do most damage there, but I quake all over when I contemplate the fact that they are keen on yew, as we have masses of yew hedges. The adults also like a bit of rhododendron, hydrangea, primula and strawberry, too.

You will have gathered by now that the vine weevil, after having a good leaf chomp, begets evil grubs, which then eat plant roots. If you do pick up a plant and find it rootless, empty the compost out on a newspaper, should you possess such an archaic but useful item, and poke it about a bit to see if the grubs are there. If the plant is outside in the ground, you must obtain a spade or trowel and search around in the dirt. If you find them and have an active bird table you can then put them out to supplement the bird food. I leave other possible ways of terminating them to your imagination and ingenuity.

But at this point, gloom must descend. Because if there are grubs in one pot or plant root, there must be the dreadful possibility that there are lots more to be found – the weevil could be roaming anywhere around the place, and is one of the most widespread of garden pests. I have some large, heavy containers in the conservatory where I once found lurking grubs. There was no way I could shovel through the compost easily and pick out the nasties, and they were present in every one. I tipped the lot out, shovelled it bit by bit into a wheelbarrow and dumped it all outside on the path in the hope that the birds would have them. I then had to add a new layer of gravel on the path and buy a lot of fresh compost, and hope desperately that every pot in the place wasn't contaminated. Thankfully, this worked – for the time being, at least. Just writing this has given me the horrors and a superstitious fear that they are still there, lurking, gorging.

I am told that if you are a hosta grower you will find it worthwhile to have a *Hosta plantaginea*. Not just because the flowers smell nice, but because the whole plant is a magnet for slugs, snails and vine weevils. A sacrificial victim, if you like, to warn you that they are about. You don't actually need warning about slugs and snails: they're always with us. But it's useful to have a vine weevil warning – again, watch for the notch.

You can, if you find you have them, set a thief to catch a thief. By now you will view fungus as the arch enemy of the stratosphere, but there are fungi that devour the revolting vine weevil grub. I'm not sure devour is right. Maybe they turn them all into little vampires, or smother them with furry fungus or rot bits off them. Anyway, unlikely as it sounds, it is called Met52 Granular Bioinsecticide and you mix it into your compost or the soil. It even does deadly things to thrip larvae, too. What could be better? Well, the challenge is that you have to catch the weevil before you're infested. It is not effective against an existing outbreak. How helpful is that? Not much, unless, I suppose, you are an ultra careful commercial grower and we hope every commercial grower is like that, don't we? Anyway, you will be glad to know that as it is not a chemical insecticide, but a fungus, it is safe and can be used with food crops.

There are also nematodes you can buy and water into the pots. *Steinernema kraussei* is time-sensitive, being most effective in late summer when the evil weevil, or

rather, its offspring, are starting to cause most damage. Another kind, *Heterohabditis megidis*, is also very temperature- and moisture-sensitive, and although I have used it, I must confess I had very little faith in it and more than my usual conviction that I would be getting it all rather expensively wrong. This was exacerbated by the fact that I mixed the nematodes up in a watering can (they don't mind this, it seems) and then had to raise a full nine-litre can up to bench height and water every pot with nematode-infested fluid. How much fluid? It's impossible to know, and watering like that, impossible to judge how much water each pot had had. What hurt as much as anything was that it was also impossible not to pour some nematode water on the ground. You can spray instead, and I wish I had known that then, but my guess would be that it would still feel like a real guessing game. For the sophisticated and methodical among us, though, you should apply it at 10 per cent of the pot volume. Therefore a 1 l pot needs 100 ml of drench – worthwhile if you only have a couple of pots, perhaps. Do the maths.

You will be delighted to hear that you can now get a less temperature-sensitive nematode called Nemasys L. Try it. With a spray it may well be doable for you. But be quick: they breed like mad – 100 can become a million in two years. Or so I am told …

⟨ WHAT TO DO ⟩

Give nematodes a try, and if you're scared you could add Met52 to your compost

Hunt them, especially in the evenings and at night. It is a sort of triumph when you find and squash one

Create little moats around all your plants by standing them on supports in saucers of water, as the weevils can't swim – who knew? Then keep the water topped up. They can climb walls and windows, though, so keep the plants away from any other surfaces. Or you could try sticky barriers placed around the pots

⟨ WHAT NOT TO DO ⟩

Ignore any notches on your leaves

Keep everywhere clean of any plant debris and other messiness where the adults might hide. (This is the horticultural equivalent of cleaning behind the fridge and never happens in the real world, but is, amazingly, frequently recommended)

Whitefly

There are some plants that I can't approach without turning a leaf upside down for a close examination. One of these is *Melianthus major*, and the reason is that I expect to find a little green (or creamy – they change colour as they mature) lump on the underside of the leaf. I'm usually unlucky and do find one, even if I have recently obtained the plant from a trusted supplier. That little lump is a whitefly nymph – doesn't that sound sweet? All of us brought up on 'Nymphs and Shepherds' will instantly break into song and hearken back to sunlit frolicking in the springtime. But this is not sweet; this is a very nasty pest. That lump will turn itself into a whitefly, which will not only feed on your plant, but also excrete honeydew on the leaves, stems and fruits, which can then lead to black sooty moulds, and potentially, like many sap-sucking insects, infect it with viruses or worse (if there is worse – that may just be rhetorical).

There is a silly illusion, which I am possibly still subject to, that squashing either fly or nymph will help eradicate the infestation if it's the only one you can find and the plant is new. This is naïve, a snare and a delusion. You might be best advised to get rid of the plant instantly by way of a major bonfire, but who is really that brave?

I get whitefly everywhere – they sometimes rise in clouds as I brush past a houseplant or greenhouse plant (they like warm temperatures, these nasties). This brings out the vacuum cleaner and some enthusiastic shredding of plant leaves as I try to suck up all the flying white insects. And I am here to tell you that you will never get them all. Even if you do, no vacuum will remove those nymphs from their safe, glued-on spot on the leaves. Vacuuming seems satisfying at first, but becomes frustrating and increasingly futile as the whitefly infested days go on …

I tried everything I could find, but in the end I was defeated. It was allied with its best mate, the red spider mite, and both are totally tenacious. Together they are a gardener's nightmare.

The first ominous thing you learn when you discover whitefly and begin to research it is that it has (like many others, it must be said) become resistant to many insecticides. Even organic ones and good soapy water. At least, those never did much for me, and there aren't many insecticides you can legally obtain and use now anyway (see Legislation, page 165).

There are biological remedies – but let's first be clear about the word 'remedy' in this context. It doesn't mean anything like to restrain or curb, as you might a mad dog or small child. It is a euphemism for *kill*. Dead. Passed. No more. And, in the case of a 'beneficial insect', it also means eaten – this insect will devour your pest. If you are lucky.

The idea of beneficial insects (actually *Encarsia formosa*, a species of wasp) does sound great, doesn't it? You imagine that you'll open a box and a whole herd of voracious, but totally benign, insects will storm into your greenhouse, home or conservatory and start munching like mad until all the nasties are gone. Then with a satisfied burp your insect in shining armour will fly away.

However, it's not quite as easy as that, and before long you will feel like you are looking after several hundred particularly difficult and demanding pets. I've been there.

You are told that you first need to monitor whether you have whitefly and in order to do this (having failed to observe swarms of the buggers all over your plants) you are told to get some sticky traps. These were once proclaimed as the solution to the pest, and I festooned both my greenhouse and conservatory with large rectangles of bright yellow plastic traps above or among the plants. The yellow was supposedly an irresistible attraction to the whitefly and the stickiness was their death trap.

The traps of course, look awful. Which is fair enough in a greenhouse, but not in a conservatory where you are growing plants for the joy of admiring them (and their infinite variety of pests, unfortunately). Despite the fact that you are unable to avoid looking at the lurid yellow traps every time you go in, you also find that it is inevitable that several times a day you will bend your head, perhaps to turn a leaf, looking for whitefly nymphs, and end up with a sticky trap stuck to your head, along with its cargo of dead insects. These never actually seem to be whitefly, but are a variety of other things that you feel, rather uncomfortably, shouldn't be executed at all, never mind in this gluey manner. You might just choose to skip this stage and get right on to the main action – those voracious whitefly scoffers.

But you can't use them unless it is warm! They need to be warm! What is warm? Well, not below 10°C at night and 18°C during the day. Buy a thermometer ... and a heater. And having obtained the requisite temperature, pray that the weather doesn't change its mind.

The first instalment of predators arrives in an envelope and needs attention straight away – these are living creatures. Straight away, that is, unless you have used an insecticide, because any residue of insecticide will kill your new pets. Frantic checking of the calendar follows, and the first flush of uncertainty: can you manage this delicate operation? There is already an unexpected amount that can go wrong, for which only you are responsible. You can't be blaming someone else and asking for new bugs if you have let the temperature drop too far last night.

So you get your little packets with lots of black dots in them – these are the cocoons of the baby *Encarsia formosa*. You wonder if they will manage to escape from their packet – do they need assistance? There is no indication that you should be poking at the packet to give them a way out, so it's probably best just to hope. You adorn your plants with these packets, which are marginally more attractive than yellow sticky traps.

Now your new daily treat is to inspect the packets and see if the black dots develop a hole. If they have done, your predator has hatched and left and you have babies! They will now start devouring.

Well, bluntly, they might. They should. But they never seemed to make a big impact in our house, mostly because my mother-in-law came to stay, found whitefly in the conservatory and helpfully sprayed them with insecticide.

I turned to succulents, as you will have discovered in the red spider mite chapter (see page 90). And having said that, I'm sure I will soon discover just what vile things succulents can be infested with.

❦ WHAT TO DO ❧

Inspect every new plant all over and if you find the characteristic green blob or a whitefly, throw it in the dustbin

Employ biological control before whitefly take over

❦ WHAT NOT TO DO ❧

Buy or accept a plant with whitefly, as this is often how it is brought into the greenhouse. Look for the green blob

Bother with contact insecticides as whitefly is usually resistant. Systemic ones might be worth a try but I have never had any joy

Ask you mother-in-law to stay (sorry, Jessica)

Ants

They don't do much harm, these pismires (I think I prefer the name 'pismire' to 'ant'.) The worst they do in my garden is occasionally to ruin one of those alpine plants that grow in a low, close mound. The plants can look excellent in a pot, especially one that is upraised, as mine was, in a chimney pot.

An ants' nest in a container like that will be a problem for your tidy plant, which will erupt like a slow volcano as soil from the roots is pushed up by the ants building their nest. Time to say goodbye to the plant, and if you are kind, sensitive and charitable, you should at least let the ants see the season out completely undisturbed and leave your pot a mess. If you try to destroy a colony, it will probably be rebuilt by new ants, which will establish even more nests in the area.

Some people are plagued by ants on their lawns, which is certainly a pain, especially if you like to go barefoot, and their excavated soil can bury lower-growing plants. Essential oils are sometimes recommended to make them run away – say a dozen drops of peppermint oil in a watering can. Worth a try? Probably not once you calculate how many gallons of scented water it would actually take. (I was also told that mice don't like the smell of peppermint. Our horrid house mice are brave and undeterred, though, so maybe some creatures have actually learned to love it.)

Ants are especially bad in the house, though. They seem to nest in our walls and once a year invade us when they want to take flight. They are a nightmare as they swarm about and fly everywhere – all over us, too, if we let them. I am very sorry to say that I have found no other remedy for this than spending the whole evening vacuuming them up. I suppose I could take the vacuum cleaner bag outdoors and let them fly off.

There is a particular ant species that you really don't want in your house, or anywhere for that matter, and that is the dreadful 'fire ant', which has a lethal addiction to electricity. Lethal to it, and to us if a colony comes chomping through our cables and nesting in plug sockets and thus causing fires and blackouts.

They have, in recent years, spread right across mainland Europe, but apparently have taken exception to the difficult conditions there and reputedly set off on a long protest march to London. We can only speculate about exactly why it was worth the migration, as the placards they were waving were too small for the human eye to decipher.

WHAT TO DO

Use ants as a signal that your plants are under attack – not by ants, but by whatever worse creatures are providing them with honeydew

WHAT NOT TO DO

Use electricity

Live in an old cottage with rubble stone walls, as ants love to nest in them

FIGHTING INFESTATIONS

Fireblight and Tulip fire

Fireblight

When I wanted to plant a hedge that would act as a permanent edge and screen to the garden, I had room to install a double line, one outside the other. So I planted one line of laurels (fast growing, evergreen, tough and, unlike the dreaded *Leylandii*, which causes murder and mayhem, tops off at about 6 m), and another that was a mix of holly and cotoneaster. The holly is good, and reliable in European climates – although it's not 100 per cent hardy, even in the UK. The cotoneaster, semi-evergreen with wonderful scarlet berries, grew much faster than the holly. The branches would be bowed down with the berries' weight come Christmas – and I emphasize the past tense in 'grew'. Many are now gone. I wouldn't go so far as to describe cotoneasters as a garden pest, although my husband might, given the trouble they have caused – he has been the person who has spent his time removing many of the dead ones.

The first cotoneasters we planted died within a couple of years and the diagnosis was not too difficult. It looked as if we'd had a fire – the outer wood looked brown and scorched, which meant fireblight.

Fireblight is a disease caused by bacteria, but it's no use trying to get a chemical to control it, or for that matter to try to find any cure for it. It's deadly. And what's more, it's contagious and used to have high drama associated with it because it was reportable in the UK. It's quite exciting, having a notifiable disease, as long as it's a plant that has it. You imagine official vehicles belting on to the property with their sirens going, and hefty rescuing men leaping out, demanding to know where the trouble is and rushing around ready to save you from danger. But all that is in the past. The UK has now resigned itself and there is no longer any need to wonder who to notify. Anyway, you can also identify fireblight by the cankers on the branches. A nasty one, this.

Sadly, the damage may well have been caused by our great ubiquitous hawthorn, which is also susceptible to the disease. I understand there are zones in Europe where efforts are still being made to contain fireblight, probably

because it not only infects cotoneaster but also many members of the rose family, including fruit trees. (I bet you didn't know that apples are a kind of rose. There's one for quiz night.)

You may be lucky and only find white goo oozing from the infected branches of your tree or shrub, which means you may have hopefully caught it early enough to cut it out and burn it. You should then prune the branch back by 30 cm (1 ft), or double that in larger specimens. If you do this, sterilize your tools during and after the cutting – dilute household bleach or Jeyes Fluid should do the trick – or you will find yourself spreading the bacteria round the place.

I missed the ooze stage, so removed the blighted trees, but being stubborn, slightly mad and just too upset, I removed only the sick cotoneasters. And then planted some more. These all survived the last twenty years, as did our accompanying apple trees and most of the roses. Some rambling roses vanished, however, so the bacteria may have persevered in some small way.

Now the cotoneasters are becoming problematic again and dying off in droves. I should perhaps remove the lot – and if I am in the business of offering sound advice, I would say don't plant them at all, but if you have and they look sick, remove them, every one. My punishment has been that I think this time they have also contracted honey fungus, which has its own chapter on page 148.

I haven't, even now, removed every bit of cotoneaster. They are still providing a screen, with great berries and flowers that the bees yearn for from the moment the buds appear, and so I simply couldn't bring myself to be ruthless enough. So what does a bad gardener like me do in these circumstances? Well, in the past we could have cut down the really dead plants and poisoned the stumps with ammonium sulphamate, but this is now illegal under European Union law (see 'Legislation', page 165).

So I have removed some clearly dying cotoneaster and will no doubt remove the rest as it dies off, but I confess I am hoping that it changes its mind and lives – I won't get overexcited about that possibility. You now know that if you have a shrub or tree that suddenly looks as if some careless idiot has recently lit a bonfire underneath, when you can see no sign of the fire itself, that you probably have a once reportable and still fatal disease in your plant.

Tulip fire

You'll be excited to hear that fireblight is not the only fiery plant problem (apart from, obviously, fire). You can also get tulip fire! It sounds thrilling, but I promise you, it isn't. It's a fungal infection and it gives your tulips (and occasionally lilies) brown spots, withered, distorted foliage, fuzzy grey mould and is generally miserable looking rather than fiery. But it does spread like wildfire.

There is an answer to this one (shock): don't grow tulips in the ground. Forget those horrible 'we've got three million tulips' displays – they are vulgar and should be the exclusive territory of over-extravagant local authorities, not gardens. You are going to grow yours in pots and, yes, throw them away after they flower.

This is not as decadent as it sounds. There are excellent companies out there, like Peter Nyssen, selling tulip bulbs in quantities that will give you a dramatic pot display at prices which are lower, even including the cost of the compost, than buying a bunch to take to your mother-in-law. And they will come in much more exciting varieties than the mother-in-law bunch.

Get a bulb catalogue and start getting excited. Even better, because you aren't expecting them to last forever, you can have new and different ones every year. You can think about having a dramatic display all of one colour. Or a vibrant clash of different colours – as long as they are strong rather than pale and pastely, they will give you a burst of pleasure every time you see them. Don't mix them in a pot, hanging basket style, however: keep the colours in separate pots and then arrange them to please as they arrive in flower. Then if you don't like the clash you can separate them, some at the front door, some at the back. Or you can go pale and interesting – have a succession of totally different ones and then experiment again next year.

When they are done, instead of watching them wilt and look miserable in your garden, you put them on the compost heap. And if yours are anything like mine, that is where they will defiantly flower again next year. Never mind, you can pick them if they do and keep them in a vase. And this way you pass the problem of all tulip diseases on to the kind growers who cultivate fields of them, while you simply enjoy the pleasure of looking at them.

If you do desperately want to grow your tulips in the ground, then check the bulbs carefully for any with signs of small black seed-like things called sclerotia, a sure sign of tulip fire, and discard any that have them on the outer scales, especially to avoid contaminating the soil. The ground should be avoided for three years where there has been a case of tulip fire, or dig deeply enough to bury the contaminated soil if you must reuse it.

✎ WHAT TO DO ✎

Get rid of anything that contracts fireblight completely, but meanwhile think cheerful thoughts as it may never happen

Grow tulips in pots

✎ WHAT NOT TO DO ✎

Plant something if there is a better alternative, as this can contaminate apples, pears and most members of the rose family

Leave plants that might be contaminated in the ground if there's even a hint of fireblight or tulip fire

Water the plants from overhead as it helps to spread the disease (sadly, rain generally comes from overhead)

Box blight

I visited a garden where the hedges had just been cut and scooped up handfuls of the cuttings into a bag – and then grew them on. It took a long time, but when they had rooted, grown a bit and been planted out, then grown on for several years (cultivating box is not a speedy undertaking), I had a lot of box hedging. And then I had a lot of box blight.

The blight didn't arrive early – I had years of heartfelt sympathy for those poor people who were removing their box hedges and topiary before I was struck with it. Or maybe them, not it, as there are at least two box blights – *Cylindrocladium buxicola* and *Volutella buxi* – fungal diseases that are often found together. Try saying those after a drink.

It may sound strange not to know if I had more than one type. But once you start looking, you see all sorts of things on your box that make it look sick. After a trim, at least with gasoline-driven machines, the cut leaves die and look a mess, for instance. It can get winter damage, starve, or rats can harm it with their urine.

I remember listening to a talk by a head gardener at a garden full of box. As he complacently told us that they were free of blight due to his knowledge and horticultural skills, a box expert standing next to me discreetly showed me the underside of one of the box leaves, with the telltale pinkish spores of *Volutella buxi*. Rightly or wrongly, no one said anything.

You will certainly know when you have *Cylindrocladium buxicola*. The dieback is dramatic and shocking. I once found a whole stretch of hedge, several feet long, all blackened and defoliated.

If you research box blight on the web you'll learn a lot, but it won't all be accurate. People will tell you how it spreads, but in reality we don't know for sure, so you may like to make all your visitors walk through a tray of bleach in case they are unwitting carriers, but this will probably only manage to turn their shoes white and their friendship toward you cold.

You'll be told to make sure you get plants from 'a reliable source'. How exactly you are supposed to determine who is reliable and who isn't, I am not sure. The word is that some nurseries are spraying their box plants with those chemicals that you used to employ as an amateur and are not now allowed to use, and which restrain but don't kill the fungus so that they seem fine until they have been in your tender care for a while. It might be advisable to quarantine any new plants until you feel sure they are healthy. I have no exact idea how long you would need to do this or just how most people could keep them away from other plants they might infect, since we don't know how it spreads. So I won't make a recommendation as to how long, as this is really a decision for you, as there is no hard and fast rule. This period should allow for any affliction to make itself known.

You will find lots of people telling you to feed your box, as that is bound to help, isn't it? Other people who seem to speak with real authority say that feeding is likely to lead to sappy growth that is more susceptible to disease. Better to be tough, make it suffer and grow hard.

The silliest advice for anyone living in wetter areas is to water your box plants from below. Apart from the time and effort involved in getting your hosepipe or watering can to do the wetting from underneath, I have to point out that rain falls directly out of the sky on to the top of everything, including your box plants. Fair enough if you live in a very dry place, perhaps – in which case you should water the plant well but infrequently in order to keep it as dry as possible above ground – but in a wetter environment this will only serve to keep your plant constantly damp.

This is all in the interest of keeping the plants dry and aerated – fungus seems to like damp, humid corners, as you may have discovered. John Sales, who used to be Gardens Advisor to the National Trust, grows his box hedges extremely thin – around 30 cm (1 ft) wide – so that light and air can reach all parts easily, which is certainly an example worth following. It also pays to trim less regularly so that you can get more ventilation through the plant, as pruning frequently leads to denser foliage and poorer circulation.

In terms of chemical control, there is ongoing experimentation and research, so there is hope for treatment yet. Currently the most effective chemicals (which protect from infection rather than cure it) are not available to amateurs, though I wonder how long that kind of restriction will persist as more and more products become available on the internet.

It would be nice if there were some good box substitutes, and many get recommended, but rigorous checking usually throws up some problem with any substitute. This could be the cost and difficulty of propagation, disease (the widely recommended *Euonymus japonica microphyllya* seems to come under suspicion here), tenderness or unsuitable growth (*Lonicera* is too vigorous and puts out straggly stems). Most of them are untested in the long term as hedges, and that in itself should urge caution. One option that isn't new to hedging and may be worth trying if you can afford it is yew, which can be kept amazingly small or narrow. And *Phillyreas* and *Osmanthus* may be worth investigating, too, the latter especially for large specimens.

Then there is the possibility of resistant varieties – 'Faulkner', 'Rococo', 'National' and 'Trompenburg', perhaps – but, as always, be cautious. I would avoid box if I was starting again.

✺ WHAT TO DO ✺

Inspect your plants for early symptoms

Grow thin (like John Sales and all good dieters) and hard

Go to the website of Didier Hermans, our saviour. http://buxuscare.com/en/pests-and-diseases is essential reading for anyone with any concerns about growing box

Treat all advice about box blight with caution – there is far too much inaccurate information out there

✺ WHAT NOT TO DO ✺

Keep feeding and watering – fungus thrives in damp conditions

Believe that blight is box's only enemy: there are a horrible number of insect pests after it, too

Clematis wilt

You are waiting for your big saucer-sized clematis to come into enormous, neighbour-wowing bloom, and every morning you rush out to see if it has happened yet. And on one of those mornings you will find that it has simply gone – collapsed, shrunken and totally deflated. Much as you are at the sight of it. You have discovered clematis wilt.

Clematis wilt is an infection that can occur in the leaves or stem, which wilt and turn the leaf stalks or stem tissue a browny-black. The spread of the disease is rapid, and from the leaf it will naturally spread to the stem and sometimes even kill the plant.

It seems that no one is quite sure what causes the wilt, although in 1989 Christopher Lloyd, a serious and knowledgeable clematis grower, was sure that it was a fungus, now called *Phoma clematidina* but then called *Ascochyta clematidina* – just in case you want to impress anyone with your erudition. He also told us that it could be prevented by using a fungicide and reported great success with Benlate. Since then, there is renewed doubt about wilt's cause and Benlate is no longer available anyway. The renamed fungus is still blamed, but just to cover all bases you will be told that it might well be due to environmental factors instead. Generally, though, we are told that the fungus *P. clematidina* readily affects the large-flowered hybrids whereas the smaller species are more resistant. Sometimes wilting occurs when the fungus is not present, no matter the size or species, and this is most likely down to environmental problems.

The confusion doesn't end there. You are frequently told to plant your clematis very deep. Your well-learned practice of planting so that the level of the soil in the pot is the same as the level of the garden soil is to be disregarded. Go deep, you are told, so the plant will be encouraged to make more shoots. You may well ask yourself why other shrubs and climbers wouldn't benefit from the same practice and grow more shoots if they were more deeply buried? You may well ask.

Dr John Howells, who spent the best part of forty years growing clematis and has done all the right things by being prominent in clematis societies and writing books about them (writing books is always a sign of great wisdom), tells us deep planting theory in gardens is nonsense. Indeed, if you ask people why it is good practice they will often become vague and mumble a bit. Howells suggests that you will get extra shoots, should you want them, by pruning after planting – the way you would pinch back many plants to increase the stems. The node, the little lump on the stem at the bottom of the plant where the flowers or leaves emerge from, is the area most likely to be affected by wilt, so it shouldn't be buried. Sad to say, he doesn't explain why the burial would make it more or less likely to wilt. Still, it's easier not to plant so deep, so if I were you I'd plant it the same way you plant everything else: level with the soil in the pot.

If you still want clematis, you could consider ones that are less likely to wilt. The nicest ones are the small flowered ones: *alpina*, *macropetala*, *integrifolia*, *diversifolia*, *viticella*, *texensis* (careful – bit temperamental, those), *montana* and *tangutica*. There are also some large-flowered clematis hybrids that are believed to be free of wilt, but which are not as straightforward to keep.

If you persevere with your clematis and you have one of those dreaded days when your beautiful, healthy and vigorous plant has a collapsed and miserable stem, what do you do then, besides weep?

Well, first of all, you cut that stem right down to the ground, or at least back to the healthy unstained tissue. Destroy the infected material so it doesn't reinfect the rest of the plant and new shoots will be able to grow at ground level. This is easier said than done, since clematis clings to anything that offers a grip. It is too late to mulch, although you should have done. And naturally you should have kept it well fed and watered and protected it from slugs and snails, which nibble the stems and let the dreaded fungus in – and make you wonder if it's not wilt at all but slug or snail damage.

Still, the good news is that wilt is rarely fatal. The bad news is that it doesn't stop after one incident. You'll have forgotten all about it and started admiring your healthy and promising clematis and bam! It'll do it again. Further good news, though, is that clematis does seem to have an amazing capacity to lie low. They may appear dead, even for a number of years, but then suddenly pop up again. So stay cheerful and never dig one up.

〜 WHAT TO DO 〜

Plant hundreds of clematis. This was what my husband did when he got clematis fever. Many got wilt; many didn't; many just didn't bother growing. But quite a few survived and now return every year. None of them are the big blowsy ones, and aren't I glad?

Go chemical if you can face it. It may work. In which case try spraying with myclobutanil or penconazole

〜 WHAT NOT TO DO 〜

Look after your clematis with meticulous care, planting well in nice, fertile, well-drained, moisture-retaining soil and mulching. Feed well, don't over- or under-water and keep off the slugs. (Yes, I have read such piffle offered as a serious suggestion)

Plant too deep

Dig your plants up straight away if you think they've died; clematis are tougher than you might expect

Plants that are trouble

Nuisance plants

*I*t's understandable when we try to move heaven and earth to save a sick pet. It's even understandable when we do that for a sick husband or a dodgy car. But a plant?

These days a plant, though often expensive, may be around the same price as a small bunch of cut flowers. When the cut flowers begin to wilt, we rather sadly and reluctantly throw them on the compost heap. If a plant begins to look poorly, however, we rush around performing first aid. We examine it top to bottom, poke at it, look for bugs, contemplate the climate, move it, spray it, cut it down in the hope of resurrection, write to plant agony aunts, listen to radio programmes with assorted 'experts' and generally waste a good deal of our lives trying to deal with the problem.

We also have a peculiar tendency to want to grow things that are dodgy to start with. Which would be fine if you wanted to spend the rest of your life dedicated to growing tender plants in Iceland or bog-loving plants in California. But if you have a normal life, with family commitments, a job and the odd pet or two, such antics are borderline crazy.

If you find yourself writing in desperate entreaty to a plant expert about what to do with a sick rose, try contemplating for a moment how much happier your life might be without that rose. What about a new space to fill and an end to the ugly blobs on sticks that so many roses are? Imagine winter and how much prettier the view out of your window will be with no prickly stalks decorating your winter flower beds. Beware plants like these that everyone 'loves': they tend to get overbred to please this indiscriminate audience and become susceptible to disease. Think how much easier it is to grow a healthy dandelion than a healthy rose and before you ask me for remedies for black spot, wonder about just how much trouble the plant is worth to you.

Clematis are the same – it's the big ones that are the most trouble and most likely to break your heart. Cultivate a taste for the smaller, sweeter, delicate flowers closer to a species. This way happiness, an easeful life and a reputation for exquisite good taste lie.

What about the newly fashionable dahlia? Well, it has a lot to recommend it, but whatever anyone tells you, it is tender in the British Isles. So think of it as an annual and be happy. You don't need to go digging it up and paying good money to keep it warm. Spend that money on next year's dahlia or grow something else.

Equally, you may have plants that clearly love you, that bounce cheerfully all over the place and always seem healthy and happy. This may mark you out as a prisoner of the idea that the greater the variety of plants you have, the better. As a result, your garden looks bitty and your happy plant is not doing its very best for you. Think of a starring role for it; think of massing it. It will fight the weeds and transform your garden. What more could you ask for?

Plants that need holding up

*N*othing looks quite as old-fashioned as a border of big sticks and string making a cage for plants to live in. It is so out of tune with the friendly, more relaxed and comfortable style that now delights us.

Still, there are plants we want to grow that are tall and rather helpless, and the answer isn't always to leave them at the nursery and buy something else. We need height too, in a border, or we'll get that prairie 'flat' look, which has its place, but that we don't want everywhere. I find myself planting things like daylilies and crocosmia next to paths because I like the way they fountain out, and their leaves look good from the moment they begin to grow, so I give them prominence. (It may also be because sometimes it's the only space where I can squeeze an extra one in.) But then they fall over the path and get trampled on and make our legs wet when we walk past. So what to do?

In one garden, I've made a feature of a railing at the edge. This, in black wood, adds definition and style in winter. In summer, I can shove tall growers behind it and keep them back from the path. Well, most of them – it seems right to let a few escape so they don't look too tidy, just as long as they don't impede the path. It's also possible to tie the plant stems back to the rail discreetly if they are too floppy.

Or I chop things down in late spring. This is a great way to show them who's boss. And then they grow again, suitably contrite, smaller, tougher and able to stand up by themselves. If they have tough spiky leaves, like crocosmia and daylilies, they will have blunt ends to their leaves for a while. They seem happy to grow new ones so that the bluntness vanishes. If they have the usual indeterminate herbaceous foliage and you cut them to the ground, they will come back as a fresh start. Think cows: plants survive being cropped by cows and pop up again. Imagine that you are a cow.

Another tip may be to avoid feeding them – or anything else in the ornamental garden – with fertilizers. By which I mean that mulching, especially by chopping down all the summer's growth on herbaceous plants, is good for the soil, helps keep it moist and is therefore good for your plants. It's usually enough without adding packet fertilizers, which may help make your plants grow too big – and therefore floppy. Feed things if they declare themselves starved by turning yellow when they aren't supposed to be yellow, or by looking weeny and feeble. Otherwise, lay off. And that includes manure, you'll be glad to know. No need to follow horses round with a shovel.

⤜ WHAT TO DO ⤛

Throw the plant away if it's a nuisance. Dig it up and get something new

⤜ WHAT NOT TO DO ⤛

Listen to Gardeners' Question Time

Algae

'Algae' refers to a variety of plants, and not all of them bad. Seaweed is algae, which is a good fertilizer. If you grow vegetables, and need to fertilize them, what better than to have a nice day by the seaside collecting some stinky seaweed and trying to get it into your car boot without leaking seawater all over the place? This practice is nothing new: George Owen of Henllys wrote in the sixteenth century about drift weed in South Wales; proof that this practice has been going on (and been smelly) for a long time, minus the handy car boot and plastic sacks.

> *This kind of ore, they gather, and lay it in great heapes, where it heteth and rotteth, and will have a strong and loathsome smell; which, being so rotten, they cast on the land, as they do their muck, and thereof springeth good corn, especially barley.*

But the algae you will be concerned with, especially if the weather is hot, will be either blanket weed, which is that lurid green hair-like stuff that emerges out of the depths of your pond and spreads across or just under the surface, sticking to things, or a kind of sludgy grey-green murky algae that looks rather like pea soup.

Algae can form when there are too many nutrients, such as nitrogen, in the water. This can come from having too many fish peeing and pooing in your pond and adding far too much nitrogen, or if your pond is fed by a natural stream flowing through farmers' fields, picking up fertilizer.

In general, you will get algae if your pond is in a sunny position with very little shade providing protection, with nothing to reflect the ultraviolet rays of the sun, or if there is a build up of organic matter from sludge, fish faeces and fallen leaves. Algae thrives in these conditions.

I tried a lot of things to remedy all of this in my early pond-tending days. There were a number of different chemicals available, all of which demanded that I did a lot of maths to work out the number of gallons or litres of water in our pond before measuring out the required amount of the formula. (I never remembered the result so had to work it out painstakingly every time, so note well: if you do this, write it down somewhere where you'll be able to find it next time.) They also came with lots of instructions and provisos and warnings, making me apprehensive about putting this stuff into the water at all, and then wondering if I'd done it wrong. It always seemed to be a great performance for no discernable result.

I then heard about the joys of floating barley straw on the pond – and this was in the days before everyone 'discovered' that it was the great cure-all, so there were no commercial products available. Feeling this *must* be the answer, I rang local farmers asking for a bale of barley straw. It was quite clear they thought I was crazy, but one was able to oblige. We didn't have a trailer and had to manhandle the bale into the car, then get a portion of it into the pond as the whole bale would have left the pond all bale and no water. But no blanket weed either, I guess. The whole thing was a nightmarish pain and as far as any results were concerned, a total waste of time.

Part of the problem with all the recommended remedies is that, because not all blanket weeds or other algae are the same, that which dispatches one will not necessarily dispatch its cousin in the next pond. The weed is evolving cunningly, and is determined to outwit all of our solutions and take over our water surfaces. We may yet defeat it by employing it commercially as a means of making biofuel, but I wouldn't count on it.

I think things have changed for me now in the fight against algae. I now have only one nitrogen-producing fish left in the pond, a big fat monstrous thing which, when a baby, managed to avoid my emptying of the pond and disposal of all the other fish. I decided they were eating tadpoles and other aquatic creatures I'd rather keep, as well as polluting the water with their excrement.

Not even the herons seemed interested enough to come and eat our fish for us. So think twice before adding fish to your pond. They start off fascinating and sweet and end up big, fat and ugly – and irremovable. Hence why one stayed, and now lurks fatly around under the water lily leaves. I keep the leaves from spreading beyond a third of the surface, though, as having some visible water and a means of seeing into it is part of the point of having a pool. Although some people will tell you that that is not enough surface cover to make any difference to algae growth, we only seem to suffer from a modicum of blanket weed now.

So, I have accepted that, like lawn mowing, an essential summer task is fishing blanket weed out of the pond. And having accepted that, I don't find it too much of a chore – it's quite satisfying and meditative. This would probably not be the case if we had a small lake, or even

a much larger pool. So, I spend happy hours (well, half-hours) fishing out blanket weed with a lawn rake, mostly wielded upside down. For smallish amounts around the edge, clinging as it does to the sides, I've found some cheap plastic salad or spaghetti servers do a good job, and I've also seen a toilet brush recommended and can see how that might work well, thought it might not release the fished out weed as easily as the servers. The connection with seaweed and the sea rapidly becomes apparent when doing this job – blanket weed smells of the seaside.

When I first made a reflective pool, I had no idea how to make it reflect. It was rather dismaying, the way that, although the base was black butyl, the sun glimmered cheerfully up at me from the bottom, highlighting all the creases in the liner and no doubting heating the water. After many queries and false starts, I discovered a black dye in America that's used in fishing lakes, presumably to keep birds from abducting the fish. This was a kind of state secret among garden designers at the time and was hard to source, but eventually I not only got some, but also a friendly correspondence with the supplier that went on for some years, until a bright spark of a food dye manufacturer in the UK decided to provide black food dye to gardeners for their pools.

Once you have started dying your water black you will need to go to rehab before you'll stop doing it. Initially I used it for the reflecting pool, but I then added it to two other ponds, which also became dark and mysterious, no longer bearing any resemblance to the grey-green, greasy Limpopo river that you will find imitated in almost every water feature. And in a pool with no incontinent fish or other live creatures, the result is less algae. The blackness of the dye reflects the sun's ultraviolet light away from the surface of the pond and in effect creates desirable anti-algae shade by not letting it photosynthesize so easily, while not harming any surface plants. Instead, you'll have tree seeds and dead flies to fish off the surface. You can get rid of these by spraying the pool with a hose end water sprayer, which will either sink them or drive them to the edge where you can fish them out with an algae net.

OTHER POND PROBLEMS

Herons

No fish, no problem. If you do insist on keeping fish, but they start vanishing fast, you can protect them with netting or, perhaps more easily, with a wooden trellis, which is easier to put down and remove. Employing either of these suggestions will, however, see your pond stray into the realm of the seriously ugly and inconvenient.

Plastic herons perched by your pond are a problem rather than a solution: they look horrible and some people think they attract male herons that are only looking for one thing, if you catch my meaning.

Water lily pests

*H*osing them off is the simplest and most effective method of combatting these bugs.

❧ WHAT TO DO ❧

Black up your pond and enjoy the reflections

Learn to enjoy a little blanket weed raking

Only feed your fish the amount of food they can eat in five minutes (you'll need a stop watch). Our fish seems to be quite happy unfed, though

❧ WHAT NOT TO DO ❧

Get some water snails. Yes, they will eat algae. And your plants, too

Vacuum your pond. (After you've finished the indoor housework?)

Add electricity and use pumps, filters or UV light. You are now into serious money, effort and possibly unpleasant noise

Bracken

My childhood home was originally called 'Ferndene', but the name was misleading in that what was growing so prolifically there was not what we generally think of as ferns, but bracken. (Gertrude Jekyll called bracken 'fern' and seems to have quite liked it. In a letter to a friend she said, 'Another of this year's pictures which pleased me was a large isolated group of foxgloves with bracken about their base, backed by a dusky wood of Scotch firs.' Bracken colours beautifully in the autumn, too.)

Many of us find bracken popping up in our flower beds, and it's not a welcome sight. It is a fern, of course, yet unlike most ferns it doesn't stay quietly in place looking pretty, but spreads rapidly. The rhizomes (continuously growing horizontal underground stems), once established, can go on spreading indefinitely. Yes, I did say indefinitely. They can invade from adjacent countryside or gardens and can germinate from airborne spores. It is reputed to cover 3 per cent of the UK's land surface and reputedly can even climb trees up to 1.5 m

(5 ft). And when you consider that it bullies other garden plants and can invade bare soil, it becomes clear that bracken is not just your standard weed.

It is dangerous stuff. You may be lucky enough not to have it in your garden, but you may encounter it on country walks and it's good to know the risks. Sheep ticks sometimes attach themselves to the bracken stalks and latch on to you as you wander past in your shorts. This can infect you with Lyme disease, which is serious and horrible – so at the least, if you do walk in long grass or bracken, think twice before showing your shapely legs off. Bracken is also a carcinogenic, so do not eat it (even if it is hawked as a delicacy) under any circumstances, and don't feed it to animals.

Bracken was once useful. It was employed as bedding for animals and was used on roads to create traction in the mud. It was good for thatch, fuel, in making dye and was given to pigs to produce a particular, and presumably desirable, flavour in bacon. Dead bracken can still be effective as mulch, and because it is ericaceous, it can be used composted as a substitute for peat compost.

There was talk at one time of cultivating it as an alternative fuel, but I believe the evidence that it won't stand repeated cutting scuppered this. This may be a clue as to how to get rid of it in your garden, but I suspect the cutting would need to be industrial and total. I have gardened with it present for over twenty-five years, frequently cutting it down without making much impact on it at all, except in the woods. You can weed kill it but you will need to do this regularly as its rhizomes go deep and creep back as soon as your back is turned.

The most popular method and the one currently regarded as most effective is bracken bashing. At the least, it is very satisfying. You take a large stick in spring or early summer and bash all the new fronds. This bruises the stem so it falls over. The idea is that it makes the plant lose energy growing a new stem, which you then bash, of course. Do this a couple of times a year and eventually the bracken will give up in despair.

It is such an effective method that some people even use machines to do it and, touchingly, they sometimes use horses to pull the machines. It's great to have horses back working in the countryside.

So don't think bracken is pretty and name your house after it. And don't ever try and pull it out with your bare hands in the summer. It is soft enough to do that in spring, which is why it is possible for some of us (like me) to be tempted to carelessly continue to do that as it coarsens up in summer. But it will rip your hands very badly.

WHAT TO DO

Ignore it: it's quite attractive and doesn't seem to displace everything else around it

Be careful at what time of year you decide to combat it

Cut it and use it as mulch

WHAT NOT TO DO

Forget its roots are carcinogenic and contain cyanide.

Never chew a root

Honey fungus

Have you ever seen what roots look like? They start out at a reasonable size near the plant – depending on the plant, tiny plants are likely to have tiny roots – then they get smaller and smaller as they spread further and further out. Honey fungus infects and spreads from those little rootlets, killing trees and shrubs and decaying the dead wood. Could you dig up a plant, do you think, and be sure of getting all those tiny end bits out? Not likely.

If you have a tree or shrub dying of honey fungus, that is what you are recommended to do. So you may be advised to get someone with a great big stump grinder to get rid of the tree stump with maximum disruption, expense and chaos. But you will probably be wasting your time because of those sneaky little roots beyond the reach of the grinder.

So start digging around the stump hole about 60 cm (2 ft) or so down and in as wide a circle as you think necessary. It's up to you what you do with all the soil and roots you've dug up, but you must dispose of it somewhere, because fresh, clean, fungus-free soil will be required to refill the hole once you have lined all the edges with something impermeable and expensive, like butyl pond liner, which will act as a physical block to the fungus. You can't get rid of the old soil where it may contaminate another area, so I recommend burning it.

Advice like this may help the 1 per cent of people with total dedication and a certain amount of madness if they live anywhere near trees they have no power to remove and which may shortly reintroduce the honey fungus. It doesn't help those of us who cannot contemplate such elephantine measures, especially when it appears tainted with the possibility of failure after all, given honey fungus is the most destructive of all fungal diseases in the UK.

What it does succeed in doing for me is making me feel guilty and obscurely responsible. If I really tried hard I could have stopped this, goes the thought. It's bad enough to have honey fungus without feeling guilty about it too. It seems to me that this is the result of much well-meant advice, and it's worth bearing in mind. People don't like to appear inadequate, so they don't like to say 'you can't do anything useful about this problem'. In the garden world especially, remorseless positivity is mandatory, so 'solutions' get offered that are possibly no solution at all in reality.

Anyway, having said all that, how do you know you have honey fungus? Well, for starters, the upper parts of the trees or shrubs, even herbaceous plants or bulbs, may suddenly die. In bits, possibly. I have a row of cotoneasters with branches dying off in sequence: as one branch is removed, the next goes. In the autumn, you may find some lovely honey coloured mushrooms, which you may eat as a kind of revenge. But be sure to cook thoroughly, and nibble a little bit first to make sure they do not disagree with you. Some people, like me, are quite sensitive to certain mushrooms and you may find the mushroom revenges itself on you rather than the reverse. And for heaven's sake, be 100 per cent sure it is honey fungus you are eating, otherwise don't risk it. (And don't mix it with alcohol if you're uncertain.)

You do not need to be frightened of the mushrooms themselves, though; they are a symptom, not a cause, and it's thought that the rhizomorphs spread the disease.

Rhizomorph is a good word with which to intimidate people and refers to what are otherwise described as 'black bootlaces', which will be found in the first couple of feet of soil. Or not, because to confuse things, the more bootlaces, the less damaging the fungus. You may have honey fungus and have neither bootlaces nor mushrooms. There are possibly six or seven different strains of this fungus (not all of them terribly damaging) and so diagnosis can be troublesome.

You can also crawl around the dying tree or shrub, sniffing under the bark. There may be a white thready mould-like growth that smells of mushrooms, (surprise!) if it is honey fungus.

You might well think that if there's not much that can be done, why worry about it? Well, you have a point. I believe the only thing you can do is attempt to thwart it by planting replacements that the fungus doesn't like. Unfortunately, this is not going to be straightforward either. I compared a Royal Horticultural Society list of susceptible trees and shrubs with a similar list by Hellis Tree Consultants. I discovered that the RHS consider beech, oak and holly susceptible, whereas Hellis considers them resistant. I'm particularly concerned myself about those, as I have hollies replacing my dying and dead cotoneaster (both acknowledge the susceptibility of those).

So I checked what the BBC had to say. They cheered us all up by making a short list of susceptible trees and shrubs (including cotoneaster) and then topped the lot by noting: 'In addition, almost all garden trees and shrubs and some herbaceous plants can also succumb.'

It may cheer you even more to know that a particular honey fungus, *Armillaria ostoyae* (apparently also now referred to as the humongous fungus), is, according to *Scientific American*, believed to be the largest living organism. It may also, at possibly 8,650 years, be the oldest.

❧ WHAT TO DO ☙

Dispose of anything you think might be contaminated

Console yourself that it's not likely that everything in your garden will die if you have honey fungus

Improve drainage and the general health of the plant if at all feasible

❧ WHAT NOT TO DO ☙

Destroy most of your garden and bank balance with stump grinders and (perhaps) drainage trenches to try and combat it

Look for chemical remedies – there are none

Weeds you want

Weeds don't always get, or deserve, a bad press. Over a hundred years ago, Gerard Manley Hopkins said in his poem 'Inversnaid':

What would the world be, once bereft
Of wet and of wildness? Let them be left,
O let them be left, wildness and wet;
Long live the weeds and the wilderness yet.

And we are lucky, just now, to be seeing a revival of wilder, weed-sympathetic gardening. If you have picked up an antiquated copy of this book, should such things continue to exist in the far-distant future, you may wonder what on earth I mean. You may be seeing a revival in enthusiasm for dwarf conifers, crazy paving and heathers. But here, right now, ecological notions and a general feeling of relaxation rule. And I am very grateful for it. Not just because it suits my gardening style (lazy), but because it is a look I prefer. I hate to see plants cut mercilessly away from the lawn by cruel clipped edges.

I don't like neat and tidy, and I positively hate the sight of bare soil or mulch surrounding each plant, sitting in loneliness, unable to embrace and communicate with its friends. This may be unavoidable in spring, when the plants are still small and haven't elbowed themselves up to their full magnificence, but it's unforgivable in high summer, when all should be pleasant luxuriance, a comfortable melding of plants.

In this context, weeds are not such a headache, partly because they are not allowed much of a foothold if they can't find bare soil to play in. Our first consideration here, before we decide what to do with those plants we really hate and must see the back of, is to decide just what qualifies as a weed. I know there is that smug line about plants in the wrong place – it is correct, but it's just repeated too often by people who are far too pleased with themselves. What it means is that you will benefit from considering your most energetic and ineradicable weeds and what they may have to offer.

Your thoughts might first turn to colour – if you have colour-sensitive borders you want your weeds to fit into. It's really no good having a pastel-pink border (once so fashionable that people fainted when Christopher Lloyd planted some cheerful colours) and then having creeping buttercup in it. Or any bright yellow weed for that matter. But if bright yellow is part of your look, let the creeping buttercup stay and contemplate the effect dispassionately. Buttercup is a weaving plant and does a good job of bringing a border together and undermining the fateful spotty look so many borders are susceptible to. As, indeed, does one of my husband's pet hates: yellow vetch (possibly *Anthyllis vulneraria*, or is it *Lathyrus pratensis*? Or *Vicia lutea*?).

Thinking about the vetches raises another important consideration for keeping a weed: how difficult is it to get rid of? I have never managed to completely pull out one of those vetches, though my determined and vengeful husband has. I prefer to contemplate the fact that it no doubt provides good things for insects, like nectar. Weeds are very likely, after all, to be our native plants, the sad remnants that the Ice Age and separation from the continent have left us. They are wild flowers and when we view them this way it conjures up images of all the good things in the countryside and thus we decide to treasure them. Perhaps.

Apart from considerations of colour, it is always good to step back and contemplate whether this plant that is giving you so much trouble might actually be worth keeping. I was confronted with my own shortsightedness in this regard when visiting a friend recently. She had a long, shady flagged path with a laurel hedge on one side and a statue as a focal point at the end. Between the stone flags and the laurel was ground elder (*Aegopodium podagraria*). The leaf of the ground elder, I realized, is actually a classy thing, and the look of this quiet garden space was also classy: cool, restrained and a break from the floweriness elsewhere. Except, of course, when the ground elder flowered, which is when it became apparent that it also has a beautiful flower, very similar to the much-admired *Ammi majus*. I believe that it doesn't spread very much by seed as the spreading factor is the rhizome under the ground, even a tiny particle of which will regenerate and grow if you try to remove the plant.

Having learned my lesson, I now also use this weed deliberately. I have a rambler rose covering a bank, and where the rose hadn't yet spread I left the ground elder for it to grow over. I added a row of terracotta pots into the subsequent sea of elder and the effect is so pleasing that I will restrain the rose if it attempts to grow over this area now. Again, it is a quiet, simple place, looking good both in leaf and flower. After it has flowered, I trim it down and soon the super leaf reappears. Excellent.

In late summer I have a very different weed-based treat. We have an area of remnant pasture where I have planted robust perennials and some bulbs. The perennials include montbretia (otherwise known as *Crocosmia* – I am not referring to 'Lucifer' but the shorter, wilder and more rampant one) and rosebay willowherb (*Chamerion angustifolium*). The willowherb, generally classed as a weed, is a pinky-purple colour, and looks amazing mixed in with the dramatic orange of the montbretia. Bright yellow creeps in with some ragwort and *Solidago* adds to the visual smack as you turn the corner and find it.

So, my point is this: before you drive yourself mad trying to get rid of any plant, consider it from all perspectives. Too many people have spotty, disturbing gardens full of too many different plants. Something that is determined to live with and love you may offer a much-needed element of unity. Or you may find a way to use the hitherto unwanted item as a star player in a new design in your garden.

✎ WHAT TO DO ✎

Consider how hard it will be to remove the weed

Think about what it can do for you if you get friendly with it

There is nothing good to say about cleavers (Galium aparine). If it was really good to eat, Tesco would sell it. But you could give it away in quantity to your worst enemy, having persuaded them that it is edible (it is, if you cook it)

✎ WHAT NOT TO DO ✎

Bindweed is a horror, never let it in unless you love the flowers

Weeds you don't want

When I started on my garden, all I had were presumed weeds because I had to clear most of two acres of field before I could do any real gardening. Having now developed a greater respect for old fields, I might be more cautious, but it's mostly done so there's no going back.

How did I do it? Well, if you are one of those really good people who never use any chemicals in their garden other than H_2O, please look away now. I did find glyphosate invaluable, together with old carpet, newspaper, grass cuttings and bark. Once I'd discovered the theory of mulching I threw my spade away in delight. (Yes, I had thought I would have to dig the whole garden by hand, and it did start out that way.) These were my mulching tools, with which I had varying degrees of success.

The old carpet was great as it was heavy and stayed in place. I got a delivery for free from a local carpet shop that wanted to get rid of the offcuts from carpet laying.

Beware, though: occasionally I forgot it was there and sometimes now when I try to plant something I find an impenetrable layer of carpet about 30 cm (1 ft) down. That's archaeology happening before our eyes (and spades), and I would have done best to insist exclusively on pure new wool, which might have rotted.

Newspaper was not bad, but had to be weighed down and would then self-shred and blow around the garden. It's rapidly becoming obsolete anyway; I wouldn't recommend it. Grass cuttings are invaluable and I still use them this way wherever bare earth appears. The only problem being that most bare earth appears in spring before you have much in the way of grass clippings. It is a nice lazy option, though, because you can take the clippings just where you need them and dump them on that dreadful, weed-inviting bare soil.

On a side note, in my experience you don't need to worry about so-called 'nitrogen depletion'. I think it's theoretical and doesn't actually happen, or not so I'd notice (your garden may be different).

Much worse is not mulching. I started one new part of the garden with the idea that I wouldn't mulch so that the plants I treasured would self-seed. Usually, I never get enough seedlings from desirable plants due to mulch. Instead the result was self-seeded weeds everywhere, and for the first time I actually saw evidence of plants looking starved in my garden. Be warned.

I should add at this point that I have discovered that some people *like* weeding. So this chapter is not for them. When I asked my mother-in-law about all the patches of bare soil she keeps around her plants, she assured me that she enjoys weeding, so she likes the bare soil. Fair enough. That's the person you need helping you in your garden. She has even valiantly had a go at ours before now, and that's dedication.

But, unlike my mother-in-law, you are probably not trying to clear a large area for unwanted plants. You are gritting your teeth and determinedly putting up with your predecessor's efforts in your garden. So you just need to know how to get rid of individual plants. Well, the obvious option, and that which the weeding lovers do, is to pull them out. Get hold of the green bit and pull. You are probably like the rest of us and inclined to do that when you wander round the garden. Especially when you are with someone who should really have your undivided attention. The dreaded question then is what to do with the bits you are now clutching in your hands? Stuff them under a hedge. The hedge will benefit. Or look for another tidy hiding place – remember, a garden is essentially a kind of stage set, with all the display on the stage itself. There are always the wings, the places where no one can see unless they are especially privileged, in this case by bending down and peering under things or looking round unlikely corners.

If this gives rise, as it does in our house, to domestic disputes, you might consider putting attractive weed containers around the place. I used to have weeding baskets – I'd buy wicker baskets and paint them a great colour (not green ...) and then place them strategically around the garden. Then you always have a dropping-off place for your weeds, and because it is purposeful and tidy – and the basket is itself attractive – you can weed happily wherever you are, knowing you won't spend the rest of your time wandering round with droopy foliage in your hand.

What do you do when the green bit breaks off in your hands, leaving a happy root chortling at you from its home in the soil? Well, you could, once alone, get a trowel or one of those useful weeding tools with a forked end and try digging that naughty root out. Or you could wait patiently for the plant to reappear, laughing at you, and then squirt it with glyphosate and laugh back. As long as it doesn't then immediately pour with rain. If the weed is hiding itself in among more desirable plants, you can use an empty plastic water bottle with the bottom cut off as a funnel. Push the bottle over the plant and squirt carefully into it. Then it may be best to leave the bottle in situ over the weed both to keep the inevitable rain off and to make sure the leaves wet with weedkiller don't come to rest on your desired plants when you remove it. I confess I'm far too impatient for all of this, so I can only say it makes good sense to try it.

In the past, we could afford to be more ecological, although some measures are now illegal. These days if you used household products, you might risk being arrested. Salt is one example, which does the same nasty trick to weeds as it does to slugs. It's shockingly bad in the soil, though, as anyone who has suffered the effects of road grit landing on their garden after cold weather will know. Surgical spirit or bleach will kill plants, but really, would you rather use them than products specifically designed for the job, even if you could? How well tested have they been for what they might do to your tomatoes or your soil?

Boiling water is good, as long as I give you a health warning (sadly, it doesn't actually come with one). Do not drink it (until it's cooler). Do not pour it over your legs instead of the plant. Or, indeed, your shoes. Or other people. You may use any you have left to make a cup of coffee, though. (Coffee is better made at slightly below boiling temperature, which your water will be by the time you get back indoors.)

You can use vinegar, too, if you are going to risk all. Unlike boiling water, you can use it in a sprayer and cover large areas. If you're posh you could use balsamic or maybe even lychee vinegar. It's not likely to kill the root of the plant, however, so your soil will be good with French fries by the time you've finished.

One thing that you really do have to accept is that weeds will always find a way. They'll be back. And some you may never be rid of, no matter what you use. Ground elder and bindweed are the very devil to eradicate and I have never had the patience to try. I accept that I have to cut the grass regularly and I guess we have to accept weeding as a similar repetitious activity. If desperate, remember that no plant will survive if deprived of light and air for long enough. You can go and chop it vengefully to the ground regularly until it gives up in despair. To be replaced by a different weed.

❧ WHAT TO DO ☙

Remember that weeds are fashionable. People have even been known to plant bindweed to grow over arches in their garden (it didn't take)

Try the cut-off bottle technique if you insist on using weedkiller and have weeds growing close to more desirable plants

Use attractive weed receptacles to ensure you can always confidently pull a weed up

Accept weeding as a permanent task, eased considerably by mulching and a garden's maturity

❧ WHAT NOT TO DO ☙

Let weeds flower and set seed, although it's true that not all weeds spread by seed

OUTWITTING HUMANS

Legislation

*T*he large, heavy feet of European legislation have stomped all over our gardens in the past few years, with some very odd effects.

You may assume that the disappearance from the market of lots of chemicals that once were useful to gardeners is because they were all dangerous to people and other nice creatures, and so you may feel grateful to the European Union Regulation (EC) Number 1107/2009 for looking after us. But it's not that simple.

Companies that wish to sell chemicals to gardeners – insecticides, weedkillers, fungicides and so on – have now to put their products through very expensive testing and then submit them for approval. In the great scheme of things, horticulture is a small market, and it is frequently not worth a manufacturer's while to bother attempting to get a product registered. So chemicals that may be totally harmless have been withdrawn and have become illegal to use as well as sell. All that may sound very good, suggesting that chemicals are all carefully monitored and at least if some are now illegal we must be safer. Nothing is ever that straightforward. Lobbying and special needs have kept some wicked chemicals available, and because they aren't designated as illegal, they sometimes get promoted as safe (although not, to my knowledge, in the UK). Legal is not actually the same thing as safe. There is still dispute about the use of neonicotinoids, which probably kill bees but are only under a temporary ban.

And the thing that shocks many people who see the legislation as supporting organic methods is that various controls that have been regarded as organic or which involve common household products as alternatives are now also illegal. The European Union finds that it is always easier to forbid than permit, and if something isn't on their list, specifically designated for the use you are after, it's illegal to use it that way. (Not just to sell it for that use, but to use it – and that means you.) So people proposing to surround a slug's dinner plant with gritty coffee grounds to discourage the slugs from having a feast will be breaking the law. However, if you are just using the coffee grounds as mulch and so not as a pesticide, you are probably all right. Probably. *Any* chemical that is used to control animals in any way is technically a pesticide and legally has to be extensively tested for effectiveness, environmental safety, operator safety and the safety of how it breaks down.

The madness extends all over the place. As you can see from the coffee example, it has now become illegal to use some chemicals for one purpose but not another. Ammonium sulphamate, for example, may be used as a compost accelerator but no longer as a weed killer.

The same sort of thing happened to what are sometimes termed heritage seeds – vegetable and fruit seeds that had been grown for generations had to be withdrawn from sale because the cost to the seed companies of registering the different varieties was simply not worthwhile.

The object of legislation was to stop unscrupulous seed-sellers selling a seed under the wrong name – hence the stringent tests (the Distinctness, Uniformity and Stability tests, under the Plant Varieties Act 1997) to clarify which seed was really which. So we were at risk of potentially losing seeds that seed companies thought weren't worth testing until gardeners' outcries produced voluntary organizations that swap seeds among keen and concerned growers, avoiding incriminating sales. This effort may someday save a critical vegetable for us if a popular variety becomes unviable due to disease or pest by maintaining genetic diversity. Remember potato blight.

These problems continue and flourish as happily as weeds. The latest one is an attempt to stop the proliferation of invasive plants, which looked as if it would have the unintended consequence of making it illegal to grow rhododendrons because *Rhododendron ponticum* is damagingly hostile in our woodlands. And don't even ask about plant breeders' rights under Regulation EC 2100/94.

All this creates a minefield for gardeners and for people attempting to write transparently and honesty about what to do about various garden problems. This is a clear warning, therefore, that some suggestions might lead to your arrest by the Itinerant Garden Police Inspector. So be careful if you spray greenfly with something you shouldn't. You have been warned.

✎ WHAT TO DO ✍

Don't worry if a product you were happily using disappears from the shelves – if your plants are looking well, move on to another product

✎ WHAT NOT TO DO ✍

Squirt any old thing over your plants without proper consideration of what it is you are using

Edges

I have never tried to suggest that gardening would be easy – what would be the point if it was? It requires trial and error and takes a lot of time to see the fruits of your labour, which in my opinion makes the whole experience all the more rewarding. But we often don't make it easy for ourselves, either. Frequently our greatest problems in the garden are of our own making – our choices of what to plant, how we want our garden to look and how we go about doing it. Rather than outwitting our garden foes, we outwit ourselves. For instance, edges …

Nothing is quite as depressing as seeing segregation in the garden. Grass this side and, rigorously separated on the other, plants. It doesn't look very friendly, and in order to keep the edge neat and tidy, the plants have to be continually policed to make sure they don't intrude onto the grass.

A special edging tool is required and many hours spent chopping away at the edges. The lawn or path gets slightly smaller all the while, of course, as you endlessly prune. Then when you've done all that and removed the debris, you have to trim any sticking-up grass with another special tool. I know. I have one, although I've not yet used it. (I did, thankfully, get it in a bundle of second-hand tools in an auction. Auction sales of house clearances are one of my best tips for garden equipment if you are just beginning. But leave the edging stuff behind. And the hoe. If you tickle the earth with a hoe, it will laugh a weed.) I suppose you could use a strimmer to tidy the grass, but you'll likely end up scalping some bits.

In order to get it right, you should get a plank of wood to cut your edge against, or if you have curves or – heaven forbid – wiggles in your lawn edge, a piece of rope or hosepipe. These can get a bit lively and can be difficult to tame, so you could try pinning each down with tent pegs, as long as you don't stick them through the hosepipe.

You may do all that and glow with pride that you have the perfect edge, but you haven't. I bet your edges don't slope by the recommended twenty to forty degrees into the border. This helps protect the grass roots from drying out and going brown, yet can require an awful amount of painstaking work.

What if your edges are not against a flower bed, but a path or paving on your patio or terrace? That should be easy: you can mow right over the edge of the grass and it will look nice and tidy. Unless they are at different heights! You don't need me to tell you that disaster awaits there. You may be back to the edging tool and hand trimming. If the difference isn't too acute you can cheerfully mow and get by, adjusting the height of the blades so as not to scalp the lawn or your hard surface. If you do that, you will hear about it: the blades will screech loudly as they blunt themselves on the stone.

That is not an end to your troubles, though. If the weather gets hot in summer, the hard surface of brick, concrete or stone that the grass is butting up to will retain heat all night long, drying it out and – you guessed it – turning it brown. If you are mad enough to want to take care of this too, you need to install insulation between the grass and the hard surface – wooden planks are one option. How do you make elegant curves with planking? All this demonstrates that gardening can lead to madness. If you find yourself worrying about any of these things, a large drink and a long holiday are advised.

If you are desperate to keep your plants from growing over the edge of your lawn, you can build a little wall of brick or stone. If this isn't flat, but is rather a small wall, then you have a new edging problem: trimming the bit of grass that abuts your new wall. This is especially fun if it's stone – tidying the grass around all the little ins and outs will provide you with hours of fun with the scissors.

Some people edge with a hedge. I'd say that works well, because the hedge itself will curtail your grass. This does not satisfy some of us. Some of us like a neat line of soil between the hedge and the grass so that we can do more edging with our tool and weeding in the bit of soil.

It is, perhaps, possible to reduce the edging work and keep all tidy by restraining the grass edge with a metal strip designed for the job. This will see to your curves and angles very nicely and leave you only the job of cutting the flat surface of the grass neatly. However, if you have a garden that undulates, there are no undulating metal strips to aid you. For metal edging to be really effective you will need to flatten everything.

So, you may well wonder what to do. Well, bear in mind that the neat and tidy garden with meticulous edging is about as unfashionable as it is possible to be. Which does also mean that any minute it will again become cutting edge (pun intended) and we will be back to trimming our edges. However, I generally let the plants and grass say hello and mingle, as I like a natural look, and it does work. My favourite border is a long chain of *Alchemilla mollis* and blue geraniums, which curves elegantly along the edge of our lawn. On the other side of the garden is a retaining wall with a deadly rough bottom where it meets the grass. I used to strim that section, which was never 100 per cent tidy, and cut the plants that were hanging elegantly over the edge of the wall in a straight line like a pudding-basin haircut. (Well, someone did.) Latterly I have just mowed as close as we can to the wall and let the edge go wild. That's been fine – it's like a mini meadow.

I mow with a ride-on mower. Originally, when I designed the garden, I edged with the lumpy stones we dug up, which worked out well, although it was a little fuzzy around the edges. But catch the wheel of your mower on one of those stones and you know about it: bent axle and a nasty shock. It will be fine with a push mower. I am in the process of removing every one of those stones. And now the borders are established, I can leave them to be constrained by the mower and an occasional strimming husband.

⧼ WHAT TO DO ⧽

Trust that au naturel is absolutely not a bad thing

Try metal strips as edges if the ground is flat

Edge with robust plants that can restrain the grass

⧼ WHAT NOT TO DO ⧽

Attempt to make all your surfaces level with each other in order to be able to mow them clean, while ensuring that you also have an angle at the edge to stop the grass drying out

Forget to clean and sharpen your edging tool

Experts

A wise woman once said, 'If you consult enough experts, you can confirm any opinion.' This is certainly true in gardening, where everyone has an opinion, and they're all conflicting. So, how do you know who to trust when you're looking for advice?

First, pick your expert. This is no simple business in the gardening world. The media – written, audio and visual – often randomly assign the title to endorse their choice of spokesperson. Which may have been simply anyone they could get hold of who could write, speak clearly, fit a certain profile or look pretty. Should you tweet about snails in a sufficiently controversial and entertaining way, you might find yourself being the light relief on the *Today* programme as a 'snail expert'. It's often a meaningless title as applied to garden personalities, writers and the rest, so always approach your choice with caution.

If you have a garden problem, you may find gardening authorities more of a pest than a cure. It's not an easy field to become truly specialist in: recent scientific studies are sadly lacking in horticulture for amateurs. So don't expect help from the academic circuit often. Your snail expert is therefore asked to opine on anything horticultural, which is an indication of the level of expertise you might be dealing with.

So, when people start holding forth, you need to ask what experience they are speaking from, which often turns out to be only their own, in their own garden. Experience from a clay-ridden plot in the south-east of the UK is hardly going to tell you what to expect in an acid loam in the north of Scotland. Has your chosen expert consulted many reliable gardeners from many other sites and situations before divulging their wisdom? Well, you may do just as well leaning over your fence and asking your neighbour, who at least has the virtue of having dealt with a roughly similar situation.

Gardening experts who have become gardening celebrities should be treated with care. They may be too busy signing their new book to actually kneel down in the soil and garden. Filming regular horticultural shows is time consuming at just the point in the year that most people feel a need to be working in their gardens, so recent practice may be lacking.

A car mechanic may deal with an endless variety of cars in different states of repair and so may ultimately develop something you may fairly consider expertise. But different types of gardens don't get seen by experts the same way mechanics see cars, and even if they did you'd still say

they couldn't be spending enough time in them to get to know the true effects of the weather, the seasons and the sheer awkwardness of the natural world. A spanner may be wielded with more accuracy than a spade.

Of course, your garden expert may have taken a horticultural course. You should ask how long ago – and you may wonder where the knowledge that is being taught came from. A case in point is the current teaching on trees. When I first learned about gardening, this is what we were told about planting trees and shrubs: you started by digging a huge hole. Having then stuck your tree into the aforementioned pit, you needed to set it at the right height. This was not simple if you got the tree bare-root and had carelessly and hurriedly 'heeled it in' a rapidly dug trench until you could plant it properly. Whatever indication on the tree trunk of just where the soil came to previously is likely to have been lost after time in a trench, thus embarrassing you when you come to dig it up and want to plant it. So you had to guess.

You would then stake the plant, grimacing at the knowledge that you were hammering a pointy stake right through the poor roots of your tree as if it were a vampire. Your stake would be upright and parallel to the tree trunk (leaving it no room to expand unless you remembered to remove the stake in time, along with the strangling tie). You'd then wrap a tree guard around it to protect it from being chomped by passing vermin.

Some of this may still apply – you could do an interesting test along the lines of 'How up to date are you?' with this example. The recommended angles of the stake have varied during my gardening life. And indeed the types of stake appear to have proliferated: you can use a single stake, a double stake, an angled stake or guying. Or no stake at all. I think things may be moving in the latter direction unless you are planting a large tree. Confused? Quite. The idea now is that wind stress helps a tree develop the kind of root system that will see it through a storm.

And enriched soil? Well, I never used to enrich the soil – it was too expensive and difficult. I remember cheerfully suggesting to an expert that it gave them a tough start and would make them into survivors. He responded that I would be unlikely to take that attitude towards a baby: more likely they'd do well having a well-fed and cared-for upbringing, so why do differently for a tree? But the most recent expert advice now supports me. You are advised to backfill with your own garden soil, having washed all compost off your new plant's roots. The reasons are complex, and Linda Chalker-Scott's blog is very helpful for finding out more – see http://blogs.extension.org/gardenprofessors/2009/07/23/introducing-linda-chalker-scott/. Or see her book, *The Informed Gardener*.

So, what to do about experts? Pick specialists who are both sceptical and experimental: apart from anything else, they tend to tell you what evidence they have for a particular practice and that is always worth having. For example, as well as Linda, try Charles Dowding's website at http://www.charlesdowding.co.uk/, or his books. If, alternatively, you are in awe of the RHS, note their advice on snowdrops: 'Pruning and training: there is no need to prune or train snowdrops. Allow the foliage to die back naturally.' Also, remember that plants like to grow and they will often do that in spite of you.

We do tend to believe advice that we approve of, however. We have our own preferences and preoccupations

– with the organic, or the chemical, perhaps – and often they mirror our personalities: what is the laziest way or the tidiest? (I'm not sure what you do when you've noted your prejudices and assessed the weaknesses you are liable to have in your own critical faculties, but it's best to be aware of them.) You may like to choose experts on that basis, and therefore you will be more inclined to follow their advice. Or that they say the 'right' kind of things. Or look pretty or enthuse dramatically on TV. But then, if their advice doesn't work, remember where we came in: 'If you consult enough experts …' – and go and find another one.

⤜ WHAT TO DO ⤛

Value your own opinion and experience: there are people who experiment and thereby save you having to do it – but it's not a bad approach for you either

Talk to your neighbours. They may know some useful things about local gardening conditions, but do add a small pinch of salt, as they may know less than you do

⤜ WHAT NOT TO DO ⤛

Trust anyone blindly. Including me

Spraying

There will be bad things discussed from the beginning in this section. Skip these pages now if you are easily offended. So, weedkilling.

If you are going to be a very bad person and spray things, you'd better do it right. You're going to kill a nasty weed with glyphosate (I told you not to look), so get out a sprayer and try to work out what a millilitre is. You will then probably wonder whether a liter is the same as a litre and how it relates to a pint. You will fondly remember quarts and gallons and get generally distracted (remember those shiny red exercise books with rods, poles and perches on the back?).

Half an hour later, you ask yourself how does a millilitre relate to the fluid ounces on your measuring jug? (Which you really shouldn't be using for nasty chemicals if the next use will be for your *Great British Bake Off* challenge cake.) You spend the next half hour doing your head in with the maths, then carefully splosh some chemical into your sprayer and add water. All over yourself. Remember to turn the tap off. You will likely think it looks pretty dilute in all that water and add another touch of the nasty chemical just to make sure (and because you spilled some when you were watering yourself).

This is very bad. It's bad enough to be spraying, and you will go to hell for that. But it's even worse not to do it properly. Start again by reading the label on the weedkiller and scaring yourself to death. Read all the things you mustn't do with it and when you mustn't do them. Note and inwardly digest. If it's snowing, give up now.

Then proceed by getting someone who can add up and understands metric measurements to show you how much chemical to use with how much water, and measure both out carefully with a dedicated utensil. This may require going off to a garden centre, so do try and possess your soul in patience. Who (besides the adverts) said it would be easy? And while you're at the garden centre, get some plastic waterproof gloves – ones that actually fit you. You will wear these when you go back home and start spraying everything. They will keep the chemicals off your skin and, more importantly, will make your hands sweaty and smelly and put you off this naughty weedkilling activity for good.

But if you're still in the game, you have to finish mixing your poison. Start by putting a little water in your sprayer from the measured amount, retaining the rest and not tripping up and spilling it all. Then add the amount of weedkiller your mathematical genius told you to add from your specially purchased measuring device. Close up the top and shake it all about so it's well mixed and then you can add the rest of the water. If there is too much water to fit in your sprayer, burst into tears and go get yourself a cup of tea. Or alternatively, put the whole lot into a disposable plastic bottle that is big enough for all of the liquid. Shake it a bit then decant once again back into the sprayer. You will then also be pleased to note that you have some spare left in your disposable plastic bottle. This will save you going through the whole rigmarole again when you run out, so put the top on the bottle and feel smug.

All this assumes you are just doing a little delicate spraying of a few unwanted plants. If you want to spray larger quantities you will need a big sprayer. And if you should want to spray with other things besides weedkillers, you would be advised to have two big sprayers, one only for weedkiller (clearly labelled in indelible ink) and one for other things, like (ahem) home-made fungicide or insect repellent – you know, smelly stuff with garlic and chillies in. Or commercial products that we don't like to talk about but that necessity drives many of us to purchase on occasion.

You would be best, in my opinion, to ignore those knapsack pressure sprayers with pump handles and many tiddly parts that you will lose if you ever do more than take the top off. They have a multitude of little hidden places for tiny bits of stuff to get into and block up the spray. They are evil. If you aren't the average candidate for a pressure sprayer, but like me female and under six feet tall, you will find it impossible to use the pumping mechanism to generate decent pressure. It would be hard work for anyone, and ultimately painful. Instead, you can get electric knapsack sprayers, which are a revelation after a pressure sprayer. They don't even tend to block up, but if they do, there are not as many places that require prodding with a pin to get the blockage out (prod with your gloves on, of course).

Now, there you are with the sprayer and the chemical, rotted garlic and chillies or some manufactured alternative. But how are you going to get the right amount sprayed on to your plants and not waste any?

Well, there is a way, supposing you can bear it. You put one of those litre/liter amounts of water into your sprayer and then spray the ground just as if you had some noxious chemical in it. Try and make a regular shape of wet and keep at it until you've emptied the sprayer. Then quickly, before you can no longer see where you've been and before it pours with rain, measure the area you've covered in square metres to go with the litres. You now know how far one litre or whatever you put in your sprayer will go.

For the truly dim among us, you may like to know that two litres will therefore go twice as far. And three … well, ask that mathematical friend of yours.

Then go spray. Preferably in the morning, because this will mean you have to get up early like good people do. And pollinating insects that won't like your spray (although it might depend on what the spray is) will still be in bed. The leaf spores of what you are spraying will be open and receptive to whatever it is you are offering them.

Your careful working out of how much stuff you needed will mean you won't have the problem of disposing of any surplus chemical, which is good because you mustn't put it down the drain. You certainly mustn't keep it unlabelled as this is illegal and, as you know, the police visit every garden shed every night to check.

ᘓᘓ WHAT TO DO ᘓᘓ

Remember spraying is not necessary for most people and most gardens. Find an alternative or if you have only a small area to treat, buy it ready mixed

ᘓᘓ WHAT NOT TO DO ᘓᘓ

Breathe it in. Mask up and wear goggles. And do remember to wash everything out very thoroughly afterwards

Bother with a manual knapsack pressure sprayer

Biofuel

This chapter could save you far more than whatever you paid for this book.

If you use small gasoline machines in your garden – a lawnmower, hedge trimmer, chainsaw or grass verge trimmer, for example – it is important to know that enthanol, a component of most, if not all, unleaded petrols, can damage them.

We've been compelled by the European Union to add biofuel to our petrol to reduce greenhouse gases. There is much debate about whether doing so has any measurable result and whether the organic materials that comprise the biofuel are actually increasing food prices, especially in poorer nations. On a micro level, I am also concerned about its effect on my machines. My lawnmower is a ride-on version, but that doesn't make your push mower immune from ethanol, unless the only power it uses is your muscle.

Our relatively new mower started having problems, coughing and spluttering like a long-term smoker and stalling at random. After much investigation, an expensive new carburettor was deemed to be the solution. It wasn't. Or at least, it didn't help. Then we got advice from a specialist in the maintenance and repair of small garden machinery, who told us the problem is the ethanol in the petrol.

The critical thing to know is that ethanol has a high oxygen content, it is acidic and it is hydroscopic, meaning it absorbs water from the air and makes the engine harder to start. It can rot your rubber fuel pipes, destroy your carburettor, rust copper and brass and damages plastics and butyl. It even harms storage tanks in petrol stations.

The lifespan of a can of petrol for use in a small machine is now said to be at most a month – though all these problems are mitigated in the larger tanks of cars of a satisfactorily recent date. Not so most of your little machines.

So, if your strimmer has started behaving badly, my best advice is to either use your can of petrol fast or drain your engines over any lengthy period of non-use. Added to that, get the relevant additive (look for an ethanol additive, sometimes referred to as a 'dry gas' or non-oxygenated additive) and make sure your petrol always contains the recommended amount of it.

Our lawnmower still hiccups when it's not under load, but at least it's not still being eaten away inside.

❧ WHAT TO DO ☙

Add an additive

Check the cap to the fuel tank for the recommendations on what fuel type to use

❧ WHAT NOT TO DO ☙

Buy more gasoline than you can use in a month

Leave your machine idle over a long period of time while it's still got fuel in the tank

Believe green fuels are always good for us

Plastics

Who on earth decided we need bright green plastic hosepipes? Or bright green plastic anything in the garden? Did someone think that gardens are green, so green plastic would look just great? And yet it never does. No artificial green looks at all like nature's green – maybe because nature always varies the tone, even across the space of a single leaf. And the greens of the great outdoors are rarely very shiny, whereas plastic has a certain inevitable tendency in that direction.

And just to add insult to aesthetic injury, my bright green hose came in a bright green plastic hose reel. Which has a remarkable power of endurance. I've been looking at a nasty big green circular bit of bright green plastic on the house wall for many years now, and time has not improved the look. This is typical, since everything else in plastic seems to disintegrate in no time.

Want to buy some trellis to grow your green beans up? Green it will be, but not green as we know it, Jim. It will come with lovely bright green plastic ties, too.

I went looking for a pot saucer and guess what? Those

come in dark green. Which is slightly better than bright green, but still screams, 'I am not plant coloured.'

I've also noticed that bright green hose connectors have lately been joined by bright orange ones – very subtle.

Perhaps the most absurd plastic item in the garden is the orange flower pot. Presumably it's because we used to have terracotta clay pots – I suppose they were a bit orangey. Hence bright orange plastic pots.

What colour would we prefer? Black. Black really can disappear, especially if it's not glossy. You can get black hosepipes, and I recommend them. They are fat, tough hoses – a bit of a pain to get the connectors on (tip: put the hose end into a jug of boiling water to soften it first), but it rarely kinks and it looks classy and unobtrusive. We could do with more of that.

But not all bits of plastic last forever, like my bright green hose reel. Most go brittle in sunlight, fade rather unpleasantly and then break into pieces or rot mysteriously. If they are the critical connectors and serve as water equipment, they will begin to irritate profoundly as they perish.

Those connecting pieces that marry your hose to the tap and to various bits of water equipment have a built-in fail factor, in my experience. They will begin by squirting water at you slyly and unexpectedly when you connect them all up and turn the tap on. If you don't get squirted in the face at the tap you will get squirted all over as you go looking for the leak in the hose. Especially if you find a kink and straighten it, thereby opening the hole further to have a good go at you.

I once reproached the manufacturers with the short life of their product and was reminded that you can repair them by buying the rubbery washers that are supposed to stop the squirt. And you can – there's another little job for you when you are wondering what to do with a spare hour, providing you have bought some of the little washers.

The other thing they tell me is that you must never let the hose end connectors, sprayers and other parts get cold. They don't like it, so bring them indoors by the fire for the winter. This is a great get-out clause for the manufacturers, of course – how could you prove you hadn't left it out in a freeze if you were taking up a warranty issue?

You may be attaching these parts to sprinklers, and they are both a real boon and a pest. For many watering jobs the waft of an oscillating sprinkler is good because it doesn't require you to try and shove a spike into rock hard dry ground, where you will inevitably encounter a stone as soon as you are near a decent depth. But these wafting sprinklers will one day, and sooner than you think, cease to waft. You will return to your watering site, expecting everything to be looking relieved and hydrated, only to discover that one plant has a remorseless and over-energetic water jet trained on it and clearly has had this destructive treatment for the last half hour, while all else stayed dry and wilted.

It is possible to re-set it, gently but firmly, and it will pretend to be all right again. It will start wafting cheerfully away, waiting only until you have disappeared indoors before coming to a dramatic full stop. Sometimes it will really kid you that you have sorted it and it may work away for long enough for you to forget it ever had a problem. But it is only waiting … someday soon it will play at being a fountain again.

❧ WHAT TO DO ❧

Buy anything you can in any colour but green until the penny drops with the manufacturers

Get a black hose and use black pots

❧ WHAT NOT TO DO ❧

Get bright green garden canes, even if they are rot proof (especially if they're rot proof)

Be surprised if you have to kit out with new watering equipment every year

Garden machinery

There are not a great many pests in this book that could kill you, but garden machinery will have a go. I have had a few hairy adventures thanks to the vagaries of various machines, including having a ride-on lawnmower drive right over the top of me. And another one perform a kind of somersault as I went up a slope so that it again ended up on top of me. Fortunately, that was a very old and small one, more like a go-cart than a mower, and I doubt you could readily get one as dangerous as that nowadays.

Or maybe you could. I had a ride-on mower not that long ago that had this amazing trick of running away with you downhill when you had your foot clamped on the brake. So, obviously, what you do then is pull on the handbrake in a panic. At that point panic turns into total terror, as the effect was to release all braking function and let the machine freewheel at uninhibited speed down the hill with a screaming driver closing her eyes and praying. Actually, quick and clever steering was the only way out and thank goodness none of our slopes ended at a cliff edge. You may be sure no one believed me about this, including my dearly beloved – until it happened to him. Which was an occasion of pure, vindictive pleasure. Especially since he didn't end up dead.

So, be warned: garden machinery can be dangerous. And no one was bothered when I wrote to them about it. I imagine they didn't believe me; I probably sounded mad. We might do better today in the social media age: a company might jump on hearing that their machine was not only lethal but that thousands of people had heard about it, too.

Machines are mostly just irritating. Anyone who can wind the line on to their strimmer will be a happy bunny, but the rest of us throw the thing away and use a blade. You can tell how successful an invention the spool system is from the multitude of videos and explanations online about how to replace your line. It's a total nightmare and

makes the sight of the ever-receding ends of your line when you are using it a depressing sight. Where do all those bits of plastic line go? Will archaeologists find them a total mystery one day in a strimmerless future? I did try plastic blades for a while, being afraid of what I'd hit using a real blade instead of a line. However, I bought several and none seemed designed to fit my machines or work properly. The answer has been to strim less. Who needs a tidy garden?

I have never managed to come to terms with a leaf sucker. I must add here that I know they are, as blowers, public enemy number one in America, because people seem not to mow up their leaves but to blow them around the place, annoying everyone in the neighbourhood. The noise of our hedge cutting is much worse than the occasional pick-up by a leaf sucker, which is usually used to pick up hedge clippings when they fall on gravel. On grass, clippings, like leaves, get mown up. The trick is to have help or a partner who just loves the tidying jobs, leaving everything neat and clean.

I am a very happy user of a little battery hedge trimmer. Not that I employ it much to trim hedges. I mostly use it to cut plants down after they've flowered or at the end of the season, or to do a bit of random pruning. I use it so much it gets blunt. The problem? If you enquire after a new blade you are told it is as cheap to replace the whole machine. Which is a sad but true story.

The best tip about the awful problem of pulling the starting cord on a little petrol machine and it refusing to start is to get one of the new recoil starts – these starter systems, called things like 'Smart Start' or 'Ergo Start', have actually made starting a small engine possible, even for people like me with very little pull. However, they haven't yet made strimmers the right size for me, and if I use one, I end up with the engine warming my ear rather unpleasantly. No power tool is actually made to be used by a woman, but for all that a great many of my female friends are the ones who do the work in the garden.

WHAT TO DO

Get someone else to do it

Get the latest equipment if you can afford it

WHAT NOT TO DO

Keep pulling if a small petrol engine won't start. It will flood the engine and render your future efforts void. If two pulls with the choke on followed by two with no choke don't work, throw it away. Or go and get a cup of tea before trying again

Noise

As Walter Bagehot says, 'An inability to stay quiet is one of the most conspicuous failings of mankind.' And I agree. When the police were called to restrain me from clouting my neighbour with a club hammer, we decided it was time to leave London. The issue was noise. That particular bullying neighbour riled the police, too, so that they nudged me to call them straight away if he started up again. Sadly, on that occasion, he went to bed and turned the music off.

And now, I regret to say, we are the noisy neighbour.

Noise really is a serious and underrated problem. It seemed in London that as soon as the sun shone people would throw open their windows and turn up whatever made music for them. We had another neighbour with an even more delightful habit. He would put his radio on the windowsill and tune it into *static* (seriously). Until I managed to poke the radio off the windowsill with a long pole, that is. Such bedlam really drives people quite mad, and there is little to be done about it when people are determined to be obnoxious.

I was disappointed when we first arrived in the country because I still found that noise was driving me mad. Partly because we had a neighbour with a large garden and an adjoining field and he spent every spare minute he could outside strimming the grass in the garden and all round the field, then back and round it all again. But it was also true that birdsong could stress me out – it was some years later that I learned that if you are exposed to very stressful noise over a long period you may get over-sensitized to all noise. This takes some time and patience to recover from, but things do improve, as they did for me: the neighbour had a heart attack (no, it wasn't me) and I de-sensitized.

So now it's my husband and I who are noisy, and I hate it. It's petrol machines, especially our hedge cutters – and we have a lot of hedges. Our neighbours are very patient about it, but I don't like what we are doing to them. The solution appears to be just over the horizon: battery power. People – men in particular – are inclined to sneer at battery tools and it's true that they sound as if they run on clockwork. But one of my best tools is a small battery hedge cutter – it cuts for ages, and if you have two batteries, one can charge while you use the other. The men won't touch it. Now you can also get a much larger brute of a thing where you carry the battery on your back. I think this is the (quieter) future and may be hefty enough for men to convince themselves it's not a toy but a real man thing. I want one. (They are still very expensive.)

If you are currently plagued every time you want peace and quiet in your garden, noise-cancelling headphones might be a help. It's not ideal, wandering round wearing headphones, yet the bliss of finding some quiet outside might just be worth it. And music in a garden can be bliss, if it's your choice and not someone else's. Unless you are too angry about the noise, in which case nothing much will help apart from a club hammer or a heart attack ...

I haven't personally had any reason to put dogs down as a garden pest, having never owned one, although I know their toilet habits can leave something to be desired – bitches' urine makes unsightly dead patches on the lawn. I understand the answer to that is to stagger out with a bucket of water every time, as the trick is to dilute it. Not the dog, the pee.

Dog noise, though, has driven me mad. One of the problems with noise is the spread – one relentlessly barking dog in the nearby village was audible all over the valley. All day, every day. And because it was in the village and not a close neighbour we didn't know the owners, so polite negotiation was not so easy. More recently, a neighbour has had just that problem with a new puppy and has been more considerate. He bought a collar that emits a nasty smell every time the bark begins, and I understand (fingers crossed) that it is working.

There will, of course, be smells that dogs are immune to, but it's a good start to dealing with a horrible problem.

One of the other things that is a noise pollutant is the radio. I remember when the wind-up radio was first delighting everyone, and one of its supposed merits was that you could use it in the garden. This, of course, drew applause from people who had never heard of a battery radio or of keeping quiet outside. There is a tech answer to this, too: Bluetooth earbuds. This allows you to have your phone in your pocket or a pouch on your belt, with a wireless connection straight to your ears instead of everyone else's. Perfect. And you can take phone calls if you actually have mobile reception.

What about putting up barriers to noise? There is a persistent illusion that there could be planting that would give you quiet – from traffic noise, perhaps. I believe the only thing that truly works is a bund – in other words, a great big earth wall between you and the world. It might also come in handy as a defence come the apocalypse. Dense hedges such as yew or laurel do cut noise slightly, however, even if not to the degree you'd like.

❧ WHAT TO DO ☙

Move to the country – it's still more likely to be quiet here and it's nicer. (Only don't all rush at once or it will get full up)

Try getting together with your neighbours and agreeing quiet times. The Independent heard we had tried this and thought it was so funny they did a half-page spread about us. They were under the impression that we were ignorant newcomers to the countryside, shocked by ordinary rural noise. In fact, we live in a very quiet part of the countryside and it's only gardening that usually disturbs the peace

❧ WHAT NOT TO DO ☙

Think going deaf helps

Get a dog

Rain

When it rains too much, we will likely have problems. Everything gives us problems, after all, so we may as well consider what these wet ones will be.

My husband and I recently spent a good deal of money installing drainage after a horribly wet winter had caused our yew hedges to start dying. They were possibly suffering from *Phytophthora*, fungus-like organisms cheerfully referred to as root rot, of which there are, of course, many different varieties – probably almost one special root rot per plant, just to be fair and leave none out. If you are worried you too may have root rot, the symptoms are a kind of fading away (which admittedly is a symptom we may all recognize …). Our plant, shrub or tree will begin to look sickly and yellow or it may wilt; yew and other conifers will turn brown and look dead. Trouble is, this is how many sicknesses take plants, so you may have to consider whether the ground is very wet and whether it has been that way for some time, as root rot is mainly a disease of waterlogged soil.

You might be amazed to know that we didn't dig the yew hedges up to find out if the roots were rotting, despite this being suggested as a reliable diagnostic method. Wet feet is one of the few things that really upset yews, so we installed the drains assuming it would help them and crossed our fingers, and indeed, green is returning to the hedges. We had never expected to have to do this as our garden is on a slope – but who knows what really goes on underground? Maybe we're sitting over a stream or hidden lake.

Certainly, if you suspect that you might one day have drainage problems in a garden you've just taken over, do the wise thing and dig trenches. In the trenches add drainage, and then, just in case you're wrong about the wet, or in case the climate changes, add water pipes and taps. I know you'd rather be buying plants, but these truly are words of wisdom.

After that, you really need to try and take care of your soil, which sounds deadly boring and suggests those mounds of misery, compost heaps. But forget all that worthy stuff, invented to keep Victorian gardeners from getting into mischief behind the potting shed, and mulch. You can mulch with your lawn cuttings, since naturally you won't have added any nasty chemical weed killers or fertilizers. You can mulch with your local tree surgeon's chippings or the fence maker's bark. (Don't worry about nitrogen depletion unless everything goes yellow, in which case you might add nitrogenous fertilizer such as some people impose on their lawns.)

There are many soil benefits from mulching. It will help lower the evaporation of water from the soil, helping it through dry weather, and it will protect the soil from compacting in heavy rain, thus helping in wet weather. Organic mulches get taken into the soil by busy little worms – at least I *think* that's how it happens, but I know that it does happen because mulches do vanish and need replenishing periodically. Once a year usually does it. And somehow all this organic matter finding its way into your soil will keep the soil structure the way it should be – open and able to drain or retain water at whim. So if you do find yourself with too much water in your soil, all that good mulch will help reduce the damage. OK, so it seems like magic, but who am I to speculate? All I know is that I mulch and that it has made clay soil friable and seems to keep it all diggable when I want to plant something. I mulch with bark, chippings and by cutting plants down and leaving them in situ in the autumn or spring. I never dig the soil except to make a hole for a new plant. Don't dig soil – it really doesn't like it. It interferes with all the microbes and other minuscules that inhabit your soil and which are happier to be left undisturbed by spade activity.

Oxygen, like water, needs to get to the roots of plants and needs a way in. Usually there is a lot of space for all that activity, but if really, really rainy weather fills space up, it tends to grind things to a halt and will eventually become smelly and grim. If you have mulched like a good person, your good soil structure will slow down that revolting process. Don't then spoil the good work by treading around on wet soil – leave it to dry out and recover.

And if your garden is likely to flood, pray it happens in winter when the plants are dormant. Before it floods, do remember to disconnect any electrical gear you may have installed in the garden. And then hope that it won't be flooded too long, which is until it wants to start growing in the spring. If you know your garden is likely to flood, grow the right plants. That is, not those that long for desert sands.

You may have deduced by now that I didn't mulch the yew hedges. It's not easy to do, and didn't appear necessary for many years when the yew grew and thrived. It just goes to show how you can garden for years and years and still get unpleasant surprises.

Snow

Snow is a great destroyer – it liberated us of our fruit cage in one catastrophic squash, as I mentioned earlier. Because of this, I used to go out in the snow bashing the stuff off all of the hedges and shrubs to save them ending up all bent over and spoilt. I did get fed up with this: the snow always goes over the top of your boots and down the back of your neck. So, in the last major snowstorm, I refrained from this miserable, snow-spoiling effort. There's always enough hard work associated with snowstorms without worrying about the garden. The yew hedges were splayed right open and thoroughly bent, but they recovered without showing any sign of injury later. Just saying. Your snow might be different.

⟐ WHAT TO DO ⟐

Remember rain is on the whole good and makes things grow

Take the hit and install some drains if you have a drainage problem

Mulch

Remember the joy of sledging in the snow

⟐ WHAT NOT TO DO ⟐

Dig

Leave bare soil

People

*I*t's not quite clear exactly what evolutionary advantage is gained by having small boys running around screaming and waving large sticks. But we are stuck with it – if you open your garden to the public you are expected to welcome this performance and I have made such things inevitable for myself by doing just this. Lock up your animals, I advise, as said small boys attacking them with sticks is not unheard of, and I suppose that is our evolutionary clue – the small boys are attempting to kill their dinner. The screaming is a slight puzzle in that case, as it does seem to induce the cats to run away and hide. If you do open up your garden, my best suggestion is to put your prices up to £20 for each child under twelve. If such a small child arrives, courtesy of a friend or relative, you are in greater difficulty as you will be required to smile and admire as the small boy jumps all over your precious veggies and risks drowning himself in your pond.

The best cure for that may be deflection – encourage children to stay indoors with some good technology and let them destroy that instead. It's less painful, even if it is more expensive to remedy.

Some people believe small children must be encouraged to learn to garden, but I think this has now been outlawed under the Human Rights Act as a cruel and unusual punishment. You must not deprive them of their tech. Indeed, you must provide recharging facilities as required.

Some small girls may, in depressingly gender-determined fashion, insist on picking that flower that you were hoping you might be able to obtain seed from and presenting it to you for your admiration. You must smile and offer various revolting slugs, beetles and other creepy crawlies to cuddle in a 'nature lesson', which will guarantee another welcome retreat of small children indoors.

Otherwise, the majority of human visitors to your garden are likely to feel very welcome, so much so that they decide to visit the garden when it is *not* open and you are wandering around in your underclothes. It is quite a shock to see strangers drifting past the window as you get out of the bath, or worse. And they really do this, believe me. If you are unable or unwilling to keep all entrances to your property locked, I can only suggest that you don't have your bathroom or bedroom on the ground floor.

I am experimenting with a stern notice on the gate, but whether that will help when visitors have travelled long distances, as they often do, I am not quite sure. There is no way to compel people to read information explaining when you are or are not open, and I have blithely made the same mistake myself, so we need perhaps to regard it as an occupational hazard.

Some problems are closer to home: the neighbours. They are clearly the greatest garden pests for many people and the problems seem endless. High hedges appear to keep on growing and blighting gardens despite the so-called High Hedges Act, which was in fact the Anti-Social Behaviour Act 2003, part 8, in case you ever need to look it up. The difficulty being, of course, that if speaking

nicely to your neighbour doesn't resolve the issue you are likely to have to take them to court. You are required by this legislation to try and settle the issue amicably, which seems to me to fly in the face of everything you'd expect to be possible with someone who inflicts a huge hedge on you. Or loud, unwelcome middle-of-the-night noises or a relentlessly barking dog. Will your neighbour suddenly stop having smelly barbeques if you ask nicely? Or will they stop letting their weeds grow through your fence or stop encouraging their cat or small children to come and pee in your garden if you smile pleasantly? You know you should stiffen up your lip and trot next door with a gift of a runner bean and a big smile and at least try. Good luck.

Boundary disagreements are famous for bringing neighbours to blows. Before you wander round with your little peace offering, do remember that people have been shot over who owns which bit of land. Inches go under dispute – maybe even millimetres these days – and such struggles can go on for years. No one seems to take any notice of the little marks or arrows on property plans or even of fences and stone walls. Anything can come into dispute when someone's land is at issue.

The best way to turn a potential enemy into a friend is to get them to do you a favour. Something small – a polite and slightly desperate request for a cup of sugar rather than launching straight into the boundary issue.

The idea is to make them feel well disposed towards you, and apparently if you are generous to someone it predisposes you to like them. Counter intuitive, I know, since most people would assume that you should be generous to them if you want their goodwill, but you may have noticed how often that backfires, making you feel resentful as well as foolish. It is certainly worth an attempt, asking for some sugar, since otherwise you may spend years in acrimonious dispute.

And who else do we gardeners hate? People who give us good advice.

For some reason, gardeners are especially committed to preserving methods of cultivation that suited our ancestors. If a Victorian head gardener insisted on keeping his huge work force busy at the slowest times of the gardening year by making them wash flower pots or polish their spades, then we must do the same. And not only do we wallow in nostalgic garden practices, we love to find *extra* hard work. If we can cut down all our perennial plants and cart them off to the compost heap, just to return said plants back to the border they came from once they have rotted and been turned and faffed about with, then don't dare tell us we don't need to do that. It's really mind-blowingly boring and hard work but the performance of these rituals proves that we are good gardeners. So we do not welcome anyone telling us it's unnecessary (although it is).

If we are offered guidance, the essential next step is for us to lie ('I *love* turning the compost heap'). A bit of quick thinking will provide us with a suitably absurd justification for relentlessly boring work in the garden and an account of why it is indispensible. It doesn't matter whether it's true or real as long as it deflects the advice. We do not want advice.

Yet for all that, gardeners are offered tips and recommendations constantly. You can't open a garden magazine or the weekend newspaper without seeing gardening advice. It churns out of the radio endlessly. It pops up on Twitter and Facebook and in every second garden blog. But we don't want it! We don't take any notice of it and we mostly hate anyone who offers it. We don't need help, thank you.

✧ WHAT TO DO ✧

Make your garden a people-free zone with a notice on the gate: 'Beware the antirrhinums'

Find a deserted island to garden

✧ WHAT NOT TO DO ✧

Listen to advice

Water

*E*very garden suffers from too much or too little water, water at the wrong time or no water at the right time. We have this strange system, poorly designed, where large amounts of water suddenly and randomly pour out of the sky on top of us, whether we ask for it or not. Or else, in an apparent sulk, water can be withheld for weeks or even months at a time, so that we begin to wonder what we did wrong and regret how flimsy our alternative arrangements are.

The alternative arrangements are that people with the least water have the lowest backup supplies and are always vulnerable to rationing and have to time their shower, bathe in an inch of water after a hard day in the garden and water their plants with an elaborate homemade piping and sucking system, which conveys the dirty bath water on to a single geranium beneath the bathroom window.

These people are also the ones who will inevitably get wet in a drought. They resort, probably illicitly in the face of a hosepipe ban, to midnight watering. You will understand by now that becoming a gardener is not, in spite of suggestions by the popular media, a harmless hobby for the challenged and middle-aged. For previously law-abiding citizens it is an introduction to the margins of law-breaking and delinquency. The threat of chemical inspections of the potting shed and helicopter surveillance of suspiciously green lawns in a drought make the mildest dabbler liable to acts of astonishing criminality.

They go out on a moonlit night with that treacherous piece of kit, the hosepipe. They then attempt to join up random bits of hosepipe to other random bits of hosepipe with odd bits of ill-fitting plastic connections in order to make the hosepipe reach where no hosepipe has ever reached before. They stumble their way back to the tap to get sprayed all over by a violently disconnected hose connector. Or, if amazingly lucky, only sprayed in a few unexpected places by a hidden hosepipe leak. Perhaps they find, when they reach the operating end of their hosepipe bits, that nothing is happening. It is as if no tap has been turned on at all. Perhaps one of the multiple connections is leaking? Torchlight inspections follow with no result. They return to the tap – the hosepipe has blown off! Careful re-inspection before attempting reconnection reveals that classic trick of the hosepipe manufacturer, the kink in the kinkless hosepipe. Unkinking produces a sudden rush of unexpected water – and so it goes on.

At some point in the middle of all this misguided effort it will probably start pouring with rain.

And, let's face it, the plants don't like tap water even after all that trouble. The difference between the result of watering a garden laboriously and thoroughly with the aid of a kinky bright green hosepipe and the effect of a good downpour of rain on a garden is dramatic. I put it down to the fact that the hosepipe is delivering copious quantities of dilute chlorine. This is expensively added to our whole water supply so that we can then flush countless gallons of it down the loo and ultimately out to sea. Why we haven't graduated beyond drains that need endless repair and using drinking quality water for the flush is a mystery to me and one I think of every time eco warriors talk of water wars or the need to put bricks in the cistern to save this precious water. Can't we produce a usable and commercial waterless composting system for our waste?

When it does rain, the rain doesn't necessarily do it correctly anyway. I spend far too much time cutting back the branches of shrubs and trees that torrential rain has bowed down over the paths. Beware falling branches, too – it's surprising how often falling rain leads to falling branches. Water, it seems, is heavy. It can certainly bash our plants about.

But better that than a light shower when you really need rain. According to many garden 'experts', a little

water on dry soil encourages the surface roots while the ones lower down sulk. Or maybe even die, because the theory is that the plants will then be more susceptible to drought (assuming you abandon all future watering to ensure this drought) because you have encouraged the wrong roots. So, if we get a shower when the garden is dry we must rush out into the garden and start watering to top it up in case the lower roots think they are being neglected, seeing the upper ones splashing happily about, and give up the ghost. Best to get water to all parts of the plant to keep it happy.

What about the midday sun suddenly coming out after rain? Researchers now say that the idea that wet leaves are scorched by sunlight is false. It will not happen, even at midday, unless the leaves are hairy. In this case, hair holds the water droplets far enough away to act as a lens to focus the sunshine. But, if you find damage after watering plants in sunshine (as I sometimes do in the conservatory), it may be you that is doing the damage by putting feed or pesticides in the water. It won't be from the rain droplets focusing the damaging rays of the sun. It will, you may be sure, be your fault whatever the reason, so you should feel bad about it.

☙ WHAT TO DO ❧

Remember that food grown in the garden will often taste better if it's been a bit short of water

Mulch. It helps with almost everything

☙ WHAT NOT TO DO ❧

Live in a dry country

Ignore a need for drainage

Wind

Wind is the worst. Don't buy a house on top of a windy hill unless you are more afraid of flooding. You can still wander around, suitably clad or with an umbrella if it's pouring with rain. But add wind and it's vicious. Add strong wind and it's terrible, and damaging to your plants. Gale force and you may have branches drop on your head as well as get miserably blown around, so best stay indoors.

But if you live out in the country, like us, you may find you are watching the wind's activities from a powerless house: extreme weather brings those branches down onto the power lines to cut the electricity. The cure for that is a generator, but you have to get your house adapted to take the power from it. When you have, it's quite cute – although fetching the generator in a gale or snowstorm is no joke, so plan ahead. The cute bit is that you plug your house into the generator, as if the house is a machine. Then the house 'goes' again, as long as your generator produces enough wattage for you.

Wind is very bad for plants. It scorches leaves, making

them look as if they are frost damaged, and can rock a tree or plant around until it's loose in the soil and the roots are damaged. It'll also dry your topsoil out if it's not mulched, although there are probably times when you might quite like that, and it can flatten overly tall plants that you forgot to cut down earlier so they'd be sturdy enough for this weather. That said, really heavy rain often does more squashing and flattening damage than wind.

If you live near the coast, the wind will give your plants a good dose of salt, and we all know salt is not good for us. There are lots of plants it's not good for either – if you've recently moved to a coastal garden, get a list from the web of salt-resistant plants. And make a windbreak from some of them while you're at it. You may already have one – before you start chopping down that row of trees that are spoiling your wonderful view, it may be wise to wonder for a moment whether it was planted for good reason. It's very difficult to manage to have both a great view and a sheltered garden. You may have to reconcile yourself to tunnel vision. You can pretend that a focused view, preferably aimed at a nice sight in the distance such as a church steeple, is ideal and that you never fancied a panorama anyway.

You might think a wall would make your garden cosy, but you know you can't afford it and it doesn't anyway. It just makes the wind cross, so it rushes over your wall and goes mad on the other side. This turbulence on the inside of the wall is called a vortex, and is rather like a whirlpool, only with cold swirling air instead of water. You can see why your plants won't like it.

Planting an evergreen hedge instead will slow down and break up the wind, reducing a lot of its bite. Although, of course, the plants that you are thus sheltering may well now suffer from the hedge's shade and from competition with the hedge's roots. You thought gardening was straightforward?

Some people insist on growing plants that really don't want to live with them. In an effort to persuade a tropical plant that they really love it, even though it is growing in a freezing wind tunnel on the east coast facing Russia, they will wrap it up for the winter. This is not only hard, tedious work, but also gives you ugly lumps of stuff to look at until spring. The idea that the appearance of a garden in winter is unimportant is ignorant and misguided. Abandon plants in overcoats and instead grow attractive sculptural hedges (whoever said they have to be flat?) or topiary. They'll look good and quite jolly with snow or, better yet, frost on them.

There are also plants that look great in winter, such as grasses like *Miscanthus*. These are plants that also enable you to enjoy the wind – they will ripple and flow in the gusts and, in my experience, come up laughing. They really can turn a curse into a blessing.

In fact, if you are living on top of that hill, a garden full of elegant grasses, which look good for nine months of the year, may be a constant delight. Sod the wind; grow grasses.

⁓ WHAT TO DO ⁓

Don't worry. It's a great pleasure to sit by a fire with a book and a glass of the good stuff listening to the wind howling round the house. Try it yourself some time

Grow grasses

Make your fences solid. After gales, new fencing sells out fast

⁓ WHAT NOT TO DO ⁓

Build a soil mound to deflect the wind

Cut down trees acting as convenient windbreaks

Conclusions

Gardening is fun, eh? Especially with a few acrobatic and demented squirrels to entertain us and induce happy, peaceful thoughts of wholesale slaughter. And you will be delighted to learn, if you didn't already know, that there are even more horrors in the garden that I haven't had time to introduce you to here. If I've missed your favourite, you might consider that the Collins book of *Pests, Diseases and Disorders of Garden Plants* (by Stefan Buckacki and Keith Harris – recommended) offers eighty-eight separate sections on plant problems. This may not sound too bad until you investigate and discover that there are sixty-seven different types of rust and forty-six different species of aphid.

So gardening certainly has its problems. And one of those may be the gardeners themselves. We're not scientists, most of us, and none of us are totally rational. So we leap to conclusions. We have fixed prejudices that we carry with great enthusiasm and sometimes with a spot of contempt for any other perspectives. Thus, some people may insist on being 'organic' and blithely use a familiar household chemical to deal with a problem, forgetting that the botanical 'chemical' solution will (we hope) have been well tested and regulated, while the household chemical has never ever been anywhere near a plant. Though it's still fun to clear a blocked toilet by pouring 2 l (3½ pts) of coke down it (and effective).

Or you could be cheerfully indifferent to rules and regulations, instructions and warnings and then find yourself poisoning your dog. You might, like a great many of us, be inconsistent and easily led. So your treatment of pests may be random and even quite mad.

We also tend to believe what people tell us, often without checking their credentials. The closer to us they are, the more likely we are to believe them, so that something your partner says will carry a lot of weight, but something your uncle's mistress has to say might be treated with great scepticism. Or, of course, possibly vice versa, depending.

You may trust things you read in books, but not in certain newspapers. You may trust someone on television or disbelieve them on principle. And our irrationality gets supported by one of the things you and I have noticed by now: no remedy against any pest appears to be 100 per cent effective. It's very frustrating and also very strange, but it is an observable truth. The most reliable product in one person's hands becomes useless in someone else's.

It may also, in some circumstances, be due to something that is not generally given enough weight, and that is the disparities in our gardens across the whole of the country. Weather and soil conditions vary dramatically and so, therefore, do a lot of other things that rely on them. It may even be possible that your red spider mite and mine are evolving at different rates in different ways and so what suits one has another one streaking for the exit.

So, my final advice? Experiment, be sceptical of all advice and if nothing you try works, be prepared to give up and go watch squirrels climb a greasy pole. Don't be too hard on yourself if you fail – there are plenty of things you can always blame. You'll never need to give up gardening, but you may have to give up a particular plant, way of growing something or even a particular prejudice. You may despair along the way, but another day will come, the sun will shine and everything will seem possible again.

ACKNOWLEDGEMENTS

My thanks to:

Hugh Barker for asking me to write this and then putting up with the consequences, and to George Maudsley for a conscientious edit and not adding me to the list of pests.

~•~

Charles Hawes for reading it and laughing.

~•~

Sara Maitland, Tristan Gregory, Kate Patel, Kathy Buxton, Margaret Roach and Victoria Summerley for their expert and generous help.

~•~

Apologies and thanks to Jessica Hawes for illuminating remarks and help in the garden, leading to my regrettable facetiousness here.

~•~

Jeff Green and Charles Hawes (again) for keeping the garden going while I disappeared.

~•~

And all those good friends on social media who answered questions and gave such generous help – you know who you all are. Xxx